T0276151

CAMBRIDGE LIBRARY COLLECTION

Books of enduring scholarly value

Technology

The focus of this series is engineering, broadly construed. It covers technological innovation from a range of periods and cultures, but centres on the technological achievements of the industrial era in the West, particularly in the nineteenth century, as understood by their contemporaries. Infrastructure is one major focus, covering the building of railways and canals, bridges and tunnels, land drainage, the laying of submarine cables, and the construction of docks and lighthouses. Other key topics include developments in industrial and manufacturing fields such as mining technology, the production of iron and steel, the use of steam power, and chemical processes such as photography and textile dyes.

A Memoir on Suspension Bridges

Published in 1832, this was the first English textbook dedicated to the topic of suspension bridges in Britain and continental Europe. Having assisted the naval officer and civil engineer Samuel Brown in preparing plans for the Clifton Suspension Bridge, Charles Stewart Drewry (1805–81) used information supplied directly by his engineering contemporaries to give an overview of the principles and challenges involved in the construction of suspension bridges. A key reference for the early history of this type of structure, the book discusses various methods and materials, ranging across rope, wood, chain and wire. Details regarding experiments on the strength of iron bars and wires are also given. Enhanced by lithographic plates and woodcut illustrations, the work is notable for its discussion of many examples of important bridges, such as Thomas Telford's Menai Suspension Bridge and the first such construction over the Thames at Hammersmith, as well as designs from overseas.

A Memoir on Suspension Bridges

*Comprising the History of Their Origin and Progress,
and of Their Application to Civil and Military Purposes*

CHARLES STEWART DREWRY

CAMBRIDGE
UNIVERSITY PRESS

CAMBRIDGE
UNIVERSITY PRESS

University Printing House, Cambridge, CB2 8BS, United Kingdom

Cambridge University Press is part of the University of Cambridge.
It furthers the University's mission by disseminating knowledge in the pursuit of
education, learning and research at the highest international levels of excellence.

www.cambridge.org
Information on this title: www.cambridge.org/9781108070539

This edition first published 1832
This digitally printed version 2014

ISBN 978-1-108-07053-9 Paperback

A

MEMOIR

ON

SUSPENSION BRIDGES,

COMPRISING THE

HISTORY OF THEIR ORIGIN AND PROGRESS,

AND OF

THEIR APPLICATION TO CIVIL AND MILITARY PURPOSES;

WITH

Descriptions of some of the most important Bridges;

VIZ.

MENAI; BERWICK; NEWHAVEN; BRIGHTON; ISLE DE BOURBON;
HAMMERSMITH; BATH; MARLOW; SHOREHAM; PONT DES INVALIDES
AT PARIS; PONT D'ARCOLE; JARNAC; TOURNON; GENEVA, ETC.

ALSO

AN ACCOUNT OF EXPERIMENTS

ON

THE STRENGTH OF IRON WIRES AND IRON BARS

AND

RULES AND TABLES FOR FACILITATING COMPUTATIONS

RELATING TO

SUSPENSION BRIDGES.

ILLUSTRATED BY LITHOGRAPHIC PLATES AND WOOD-CUTS.

BY

CHARLES STEWART DREWRY,

ASSOCIATE MEMBER OF THE INSTITUTION OF CIVIL ENGINEERS.

LONDON:

PRINTED FOR

LONGMAN, REES, ORME, BROWN, GREEN, & LONGMAN
PATERNOSTER-ROW.

1832.

TO

COMMANDER SAMUEL BROWN, R.N.

THIS WORK IS INSCRIBED

BY HIS OBLIGED FRIEND,

THE AUTHOR.

A 3

Lᴏɴᴅᴏɴ.
Printed by A. and R. Spottiswoode,
New-Street-Square.

PREFACE.

THE great extension that has been given, within the last ten years, to Suspension Bridges, and the hold they have acquired on public attention, have begun to render them so much an object of general as well as professional interest, that the want of something like a methodical treatise on them is beginning to be felt. Accounts of the most remarkable Suspension Bridges have been published, at various times, in scientific Journals; and investigations of parts of the theory are to be met with in works on other branches of mechanical science. But, except a very short work by Mr. Cumming, and the account of the erection of the Menai Bridge, by Mr. Provis, we have no book in the language treating exclusively of Suspension Bridges. A blank is thus left in professional literature, which it has been the attempt of the Author to fill up.

His object, therefore, in the following pages, has been to collect into one volume whatever he could find interesting and useful on Suspension Bridges; *viz.* first, to draw up a connected account of the History of Suspension Bridges, followed by descriptions of the most important works of that class. Secondly, to draw from the practice of eminent Engineers inferences useful to those who have not opportunities of acquiring, by practice, a knowledge of Suspension Bridges; and to apply to this branch of engineering, rules which have been established by long practice in other departments of mechanical construction.

In a work of this character, much, particularly in the descriptive part, must, from its very nature, be compilation; and, accordingly, much has been selected from the scattered information communicated by other writers, in detached accounts, and in Papers and Reports printed in various scientific Journals.

To the writers of whose previous labours the Author has thus availed himself, he takes this opportunity of acknowledging his obligations; and as he has generally been careful to cite his authorities, his readers will know both to whom to assign the credit, and where to find the originals, if they desire so to do.* He has great pleasure, in particular, in expressing how much he is indebted to Captain S. Brown, R.N., Mr. W. Tierney Clark, and Mr. Brunel, jun., for the kindness with which they have communicated to him information on their works.

For the few opinions and rules which proceed from the Author himself, as they have no established authority to support them, so they will, of course, be

---

* A list of the works cited and referred to is subjoined.

received with doubt, and examined with severity.
The method which he has pursued in forming the
rules has been to establish some mode of calculation
on the groundwork of experiments, and on the received principles of the strength of materials; and
then to modify the formula so constructed, until its
results would correspond tolerably with the proportions adopted in practice in the best existing examples of Suspension Bridges. This method is not,
perhaps, the most scientific, but it is sufficient for
practical purposes, because the object of rules, in
practical construction, is to find results for new cases,
proportionate to those that time has stamped as sufficient in previous practice. Experience, therefore,
alone can determine how far the rules given are effi-
cient; and if, upon trial, they are found to be so,
the object of the Author will be attained.

Chancery Lane, London,
September, 1832.

CONTENTS.

SECTION I.

HISTORICAL ACCOUNT OF BRIDGES OF SUSPENSION.

SECTION II.

EXPERIMENTS ON THE STRENGTH OF IRON WIRES, IRON BARS AND CHAINS, AND STEEL BARS.

SECTION III.

ACCOUNT OF SMALL WIRE BRIDGES IN GREAT BRITAIN, AND INTRODUCTION OF BAR CHAINS FOR SUSPENSION BRIDGES.

SECTION IV.

FOREIGN BRIDGES.

xii   CONTENTS.

A
LIST OF THE BOOKS AND PAPERS

CITED IN THE FOLLOWING WORK.

W. A. Provis. Account of the Erection of the Menai Bridge, in folio. London. 1828.

Stevenson. A Paper on Suspension Bridges. Edinburgh Phil. Journ. No. 10.

Captain Brown. Account of Trinity Pier. Edinburgh Phil. Journ. No. 11.

Navier. Mémoire sur les Ponts Suspendus. Paris: Bachelier.

Seguin. Sur les Ponts Suspendus. Paris: Bachelier.

Quenôt. Description du Pont de Jarnac. Paris: Bachelier.

Vicât. Description du Pont d'Argentât. Paris: Bachelier.

I. von Mitis. An Account of a Steel Suspension Bridge at Vienna. Published in German.*

Professor Barlow. Essay on the Strength and Stress of Timber. London.

Farey. Treatise on the Steam Engine. London.

Dufour. Description du Pont en Fil de Fer à Genève. Paris: Bachelier.

Symes. Account of Brighton Pier. Brighton.

Buchanan. Report on a Suspension Bridge proposed at Montrose. Edinburgh Phil. Journ. Nos. 21. and 22.

Parliamentary Reports on the Menai Bridge.

Dr. Gregory. Mathematics for Practical Men. London.

---

* A copy is in the library of the Institution of Civil Engineers.

ERRATA.

Page 19. line 1. of note, for " in art. 28." read " in art. 29."

26. note, for " Stephenson " read " *Stevenson.* "

61. last line of note, for " art. 33." read " art. 32."

69. line 10. from bottom, for " 256 feet " read " 255 feet."

89. lines 9. and 10. from bottom, for " by coupling links " read " by *open* coupling links."

91. lines 15. and 16. for " 2 persons per square foot " read " 1 person per 2 square feet."

95. the small figure, which is an end or cross view of the bridge, is placed upside down.

102. last line but one, for " square inches + " read " square inches × ."

106. line 12., for " + 8" read " × 8."

111. line 21., for " on *the* two lines " read " on two lines."

139. line 6., for " joints," read " joists."

170. for $\left\{\begin{array}{l}\text{" The strain on}\\\text{the point A,}\\\text{for instance}\end{array}\right\} = \frac{\text{A } e \times \frac{1}{2}\text{weight}}{1\,b\,e}$" read "$\frac{\text{A } e \times \frac{1}{2}\text{weight}}{b\,e}$."

179. line 13. from bottom, for " in page 162." read " in page 167."

189. line 3., for " by the foregoing rule" read " by the standard of 70 lbs. per square foot of platform."

194. for " Then the breadth $= \sqrt{\dfrac{56000 \times 20}{250 \times 1296}}$" read "$\dfrac{56000 \times 20}{250 \times 1296}$."

201. line 5. from bottom, for " in art. 192." read " in art. 194."

MEMOIR

SUSPENSION BRIDGES.

SECTION I.

HISTORICAL ACCOUNT OF BRIDGES OF SUSPENSION.

1. BRIDGES made of ropes were known in South America before its discovery by the Europeans.

De Ulloa, a Spanish writer, describes a kind of bridge called *tarabita*, used to cross the valleys of the Cordilleras in South America.*

A cable made of bamboo, or of strips of hide, is stretched from one side of the valley to the other. It is fastened at one end to a post, and the other is passed round a wheel to strain the rope more or less tight. A basket, or sort of cradle, large enough for a man to sit in it, is suspended from the rope by two bridles or loops, and the rope being a little inclined downwards from one bank to the other, a man easily works himself across. There are two such ropes inclined in opposite directions, for passing and repassing.

2. Captain Basil Hall mentions, in his journal, having crossed in Chili a suspension bridge made of

* De Ulloa, Voyage Historique de l'Amérique Méridionale, 2 tom. Paris, 1752.

B

hide ropes over the river Maypo, near the scene of a celebrated battle fought during the war of independence, in which General San Martin, at the head of the Chilian army, defeated the Spaniards on the 5th of May, 1818.

It consists, according to Captain Hall, of a narrow roadway of planks, laid crosswise, with their ends resting on straight ropes, suspended by means of short vertical cords, from a set of thicker ropes drawn across from bank to bank. The sustaining ropes are six in number, and hang in flat curves, one above the other ; the short vertical suspending cords being so disposed as to distribute the weight equally : the main sustaining ropes are firmly secured to the angles of the rock on one side, at the height of thirty feet above the stream ; but the opposite bank being low, it has been found necessary to carry the ropes over a high wooden frame-work, and to attach them afterwards to trees and to posts driven into the bank.

The clear span is 123 feet. " The materials being " very elastic," says Captain Hall, " the bridge " waved up and down with our weight, and vibrated " in so alarming a manner, that we dismounted and " drove our horses one by one before us." *

Dr. Robertson mentions also bridges of this kind on the authority of De Ulloa, as existing in Peru before it was invaded by the Spaniards.†

3. A somewhat similar kind of bridge is used in the Himmala district in India. ‡

---

* Extracts from a Journal written on the Coasts of Chili, Peru, and Mexico, in 1820, 1821, 1822, by Capt. Basil Hall, R.N. vol. ii. p. 115.

† History of America, p. 63., book vii.

‡ Fraser's Journal of a Tour through part of the snowy range of the Himmala Mountains, p. 260. London, 1820.

" A communication," says Fraser, " is kept up
" across the Sutlej by means of that singular and
" dangerous kind of bridge, which in the hills is
" called a *JHoola*.

" At some convenient spot where the river is nar-
" row, and the rocks on either side overhang the
" stream, a short beam of wood is fixed horizontally
" upon or behind two strong stakes, that are driven
" into the banks on each side of the water, and round
" these beams ropes are strained, extending from the
" one to the other across the river, and they are
" hauled tight, or kept in their places by a sort of
" windlass. The rope used in forming this bridge is
" usually from two to three inches in circumference,
" at least nine or ten times crossed to make it
" secure.

" This collection of ropes is traversed by a block
" of wood, hollowed into a semicircular groove, large
" enough to slide easily along it ; and around this
" block ropes are suspended, forming a loop, in which
" passengers seat themselves, clasping its upper part
" with their hands, to keep themselves steady.

" A line fixed to the wooden block at each end,
" extending to each bank, serves to haul it and the
" passenger attached to it from one side of the river
" to the other.

" The *JHoola*, at Rampore, was somewhat formid-
" able ; for the river tumbles beneath it in a very
" awful way, and the ropes, though they decline in
" the centre to the water, are elevated from thirty to
" forty feet above it. The span is from ninety to
" a hundred yards."

4. The same method is mentioned by Santa Cruz,
in his *Réflexions Militaires*, for getting artillery over

rivers, by slinging the carriages to blocks running upon cables stretched across from bank to bank.*

5. Baron Humboldt, in his *Vues des Cordilières, et Monumens des Peuples indigènes de l'Amerique,* describes a rope bridge, called the Bridge of Penipè, over the river *Chambo,* which he crossed in June, 1802.

It is made of ropes, very nearly four inches in diameter, made of the fibres of the *Agava Americana.*

The length of the bridge is 131 feet, and the breadth rather more than 8 feet. The main ropes are bound round with rings of bamboo, and are fastened, on each bank, to a sort of rough framework, made of trunks of trees. (See fig. 1. Plate I.)

There are other bridges of this kind of much larger dimensions; and it is by means of such a bridge, of extraordinary length, over which passengers and loaded mules can pass, that a communication has lately been established between Lima and Quito,' after a fruitless expenditure of one million of francs (40,000*l.*) in endeavouring to build a stone bridge near *Santa,* over a torrent that rushes down the Cordillera of the Andes.†

6. There is at Eglisau, near Zurich, in Switzerland, a wooden bridge over the Rhine, which, although it cannot be called strictly a suspension bridge, partakes of the suspension principle. It is in two arches, with a pier in the middle of the river, and an abutment on each bank. An arch is thrown over each opening, made of wood beams about fifteen feet long, and ten inches broad by twelve deep, with their ends abutting against each other, tied

---

* Douglas on Military Bridges. Vol. i. 8vo. London. 1816.
† Navier, p. 21.

and bolted together with iron straps. The extremities of the arch are tied together by a horizontal brace. The platform is attached to the lower ends of vertical suspending beams, twelve inches square, placed about three feet apart, in pairs, the upper ends of which embrace the wood arch, and are bolted together over the top of it. The platform is thus really suspended from the arch, although the construction is not flexible, as it is in rope or chain bridges.

7. We are informed by travellers, that several suspension bridges, made of iron chains, exist in China.

Turner and Major Rennell * describe one built over the river Sampoo, on the road to Lassa, called *Chouca-cha-zum*. It is made of five parallel chains, with links one foot diameter, on which a loose bamboo flooring is laid.

8. Another bridge, called *Selo-cha-zum*, is described by Turner, which approximates, in its arrangement, to our modern European chain bridges.

It is formed of two parallel chains, four feet apart, which are suspended over stone piers about eight feet high on each bank. The ends of the chains pass back thence, and turn down in an oblique direction, and are bedded in the rock, each being fastened round a large stone, which is kept down by a mass of smaller stones laid upon it. A plank, about eight inches wide, extending across the river, is suspended from the chains by bands made of roots, of such length, that the path is four feet below the chains in the middle of the length of the bridge. The length of the bridge is 59 feet. The suspending bands are

* Turner's Account of the Embassy to Thibet, and Rennell's Description of Hindostan.

renewed every year, and the planks are loose, so that any part can be repaired separately. This is, in fact, a proper suspension bridge, with a horizontal platform suspended from the main chains. It is only used for foot-passengers.

There are other bridges in China of the same kind. One of these bridges in particular, over a torrent at Quay Cheu, is of great celebrity and antiquity, and is called *The Iron Bridge*, as a sort of distinction. It is the work of an ancient Chinese general, and has been a model for others.

9. It would be useless labour to seek, in the accounts of travellers, more ample details concerning these bridges ; for, although they are interesting as evidence of the great antiquity of suspension bridges, they afford but little practical instruction applicable to the construction of our chain bridges. It may be concluded, however, from the accounts that we have, that iron suspension bridges are of Asiatic origin. The bridge of Chouka is, indeed, so ancient, that the inhabitants are ignorant of the date of its erection, and attribute to it a fabulous origin.

10. In Europe rope bridges were formerly much used in war. A bridge of ropes across the Clain, at the siege of Poictiers, in the reign of Charles IX. of France, is mentioned in *Davila's Historia delle Guerre Civile di Francia*, vol. i. p. 264.

Bridges of cordage were also used by Henry, Prince of Orange, in 1631, in an enterprise against Ghent and Bruges. They were also used in Italy in the campaigns of 1742.* " But no details of their construction having been preserved (says Sir H. Douglas), they were little known in later times, until 1792, when they were used by the French."

* Douglas on Military Bridges, p. 168.

11. Figures 2. and 3., Plate I., are an elevation and plan of a rope bridge, taken from a French military work called *Aide Mémoire* (cited in *Douglas on Military Bridges*). The platform is suspended from two cables, about 6 inches in circumference, and 10 feet apart, passed over gins, or shears, fixed on each bank of the river, and strained up by tackle worked by capstans placed on the shore. From these cables, cross beams 11 feet long and 4 inches square, with iron rings at their ends, are suspended by ropes ; and on those cross beams six ropes, (K. fig. 3.) 3 inches in diameter, are laid at 19 inches apart, one end of the ropes being made fast to piles driven in the bank, and the other to a capstan. These ropes are supported on frames, or trestles, placed on each side of the river to form the abutments of the bridge. The tops of the trestles are 16 feet long and 1 foot square (see fig. 3.), with grooves cut in them to receive the six ropes ; two inch planking is laid across the ropes K, and thus forms the roadway.

The weight of the cordage, and other appurtenances for such a bridge over a river 130 feet wide, is estimated in the *Aide Mémoire* at 14,878 lbs. French weight. " When infantry are marching over" (says Sir H. Douglas) " under a front of three men, ranks three " feet distant from each other, the bridge will have to " sustain about 120 men. Their total weight, each " taken with his arms and accoutrements, &c. at about " 180 lbs., is about 21,600 lbs. ; which, added to " 14,870 lbs., gives 36,470 lbs. (about $16\frac{1}{4}$ tons), the " weight to be borne by the cables and the four " trestles."

12. This is in some sort a double suspension bridge. For the lower ropes K, are a bridge themselves, if they are held fast by their piles and capstan ; and

B 4

do not, therefore, depend exclusively on the main cables, although they are supported by them.

The construction appears, however, loose, and insufficient in strength for the load proposed.

13. A rope bridge was made in 1810, by Colonel Sturgeon, of the staff corps, to cross a broken arch of the bridge of Alcantara over the Tagus.

The rope-work to support the flooring of the bridge was made of hawsers, forming a kind of net-work A A (see fig. 4. Plate I.); to each end of which a cross beam B, was fastened, and the net stretched by tackle from the fixed beam R, lodged in the masonry of the bridge. Cross beams or joists were then lashed to the rope net, and sleepers laid along them, on which the floor was laid, composed of boards lashed together with spun yarn, for convenience in rolling up and stowing in carriages.

A tarpaulin was stretched along the outsides of the bridge to form a parapet and blind for cattle.

The platform was steadied by ropes lashed to two of the cross-bearers, and made fast to the masonry below.

The span of the arch was 100 feet.

The means of transport required for the whole apparatus were,

1 pontoon carriage drawn by 12 oxen.
8 strong cars, 4 oxen each   32
17 cars, 2 oxen each   34
                 —
Total 78 oxen.*

14. M. Navier says there is a rope bridge now in use in the neighbourhood of Brest, communicating between the sea-shore and a small fortified island.

* Douglas on Military Bridges, p. 176.

He gives also a description of a suspension bridge made of ropes, taken from an old work, but without any date, entitled, *Machinæ novæ Fausti Verantii Siceni.*

It consists of cables stretched across a river, and passed over upright piles of wood fixed on either bank.

The roadway is a horizontal platform of wood, suspended by ropes that pass round sheaves, in blocks fastened to the main cables or catenaries.

It appears to have been designed for military uses, for it is made to take to pieces readily ; and the main cables, as well as the vertical suspending cords, are passed round sheaves, to tighten or slacken them at pleasure.

15. We have no accounts of the existence of iron suspension bridges in Europe before the middle of the last century.

The first European chain bridge was built in England across the Tees, two miles above Middleton, chiefly for the use of the miners of that district. The length was 70 feet, the breadth rather more than 2 feet, and the height above the water 60 feet. It is described by Hutchinson, in his *Antiquities of Durham*, published in 1794. Mr. Stevenson, of Edinburgh, takes the date of the construction of this bridge to be about 1741.*

Monsieur Navier speaks of a chain stretched across between two rocks that command the town of Moustiers, in the *Département des Basses Alpes.* It is 656 feet long, made of rods about 2 feet 1½ inch long, and ¾ inch diameter, hooked one to the other without any intermediate links. The date of its

* See An Account of Suspension Bridges, by Robert Stevenson, Esq. F. R. S. E., in the Edinburgh Phil. Journal, No. 10.

erection is not certain, but is supposed to belong to
the 13th century. It does not, however, appear
to have been ever intended for a bridge, but is thought
by some to have been an offering to the Virgin, to
obtain her protection for the town of Moustiers,
against being overwhelmed by the rocks that overhang it. By others it is attributed to a knight of
Rhodes, who is supposed to have erected it in consequence of a vow made during his captivity in the
Holy Land. The iron is said not to be injured by
rust.*

16. We may, therefore, pronounce with tolerable
confidence, that the bridge over the Tees, was the
first iron suspension bridge constructed in Europe.

From the account given of it, it was a very rude
work, in nowise superior to the Chinese chain bridges,
and the construction of suspension bridges does not
appear again publicly to have engaged the attention of
English engineers until 1814.

17. In 1807, a project was brought forward by
Monsieur Belu, a French engineer, for establishing a
bridge between Wesel and Ruderich over the Rhine.

The masses of ice that float down the Rhine in
winter, render the construction and durability of
piers almost an impossibility. And it was, therefore,
deemed requisite to throw a bridge of one arch, at
least over one arm, the breadth of which is 820 feet.
Mons. Belu's design was to lay the roadway on a
sort of network of wrought iron chains, formed by
laying a number of chains together lengthwise and
crosswise. The chains being made of small links
1·64 feet long, connected by pins. The web was to
be fixed at each end to a mass of masonry, and the
deflection proposed was $26\frac{1}{4}$ feet, or $\frac{1}{312}$th part of

* Navier, p. 24.

the chord line. The roadway at the ends, where the curve would have been rapid, was to be supported by suspension rods, to keep it at a reasonable inclination.* Mons. Navier says the objection made to the design, was principally the slightness of the chains compared to their tension. This was a fault so easily remedied, that it is more probable the objection arose from other more weighty circumstances. The scheme is altogether, if we may judge from Mons. Navier's account of it, too crude for a work of such magnitude, and was not likely to inspire any confidence at that period, when there was no experience on the subject.

American Suspension Bridges.

18. The construction of iron suspension bridges was begun in America in 1796, by Mr. Finlay, who in that year built one about 70 feet long, across Jacob's Creek on the road between Union Town and Greenburgh.

Mr. Finlay took a patent in 1801 for the construction of suspension bridges; and it is stated in *Pope's Treatise on Bridge Architecture*, published at New York in 1811, that since 1801, eight bridges on Finlay's plan had been built.†

---

* Navier, p. 26.

† Mr. Finlay's specification describes a bridge supported by two chains, the ends of which are to be firmly fastened in the ground. He directs the suspension towers to be of such height that the deflexion of the curve shall be one seventh of the span. The two central joists or cross-beams of the platform to rest on the chains; the others to be supported by vertical suspending chains : the angle of the back stays to be the same as the angle of direction of the curve, and the ends of the chains to pass through heavy stones, and be keyed under them, with a mass of masonry or stone over them; the vertical suspending chains to be attached to those of the links of the main-chains that are placed flat-wise, by passing up through them,

19. The largest across the Cataract of Schuylkill, is 306 feet long. It is borne by two chains, one on each side, the iron of which is 1½ inch square.

Another across the Brandywine at Wilmington, is 145 feet long. It is 30 feet wide, and is borne by four chains, the iron of which is 1⅜ inch diameter.

20. Another, of considerable size and strength, is described in the same work, built over the Merrimack, three miles above Newbury Port.

It consists of one arch of 244 feet span. The piers upon which the suspension pillars are erected, are of stone, 47 feet long, and 37 feet high. The suspension pillars are 35 feet high, and are supported by wood framing. The bridge is formed of ten chains, the ends of which are carried down into deep wells in each bank, and are therein fixed to heavy stones.

The total length of each chain is 516 feet ; and at the parts which lie on the top of the suspension pillars the chains are tripled and made with short links. The four middle joists or cross timbers of the roadway rest on the chains themselves. The other joists are suspended from the chains by rods to keep the road horizontal.

There are two roadways, 15 feet broad each, and the floor is strengthened by a thick side rail or parapet. The chains are arranged thus : three at each side of the bridge parallel to each other, and four in the middle, and they are computed to be capable of bearing 500 tons. The height from the water to the floor is 40 feet : the chains are painted.

---

and being held by a cross-key over the chains; to the links that are placed upright, or on end, the vertical suspending chains are held by a fork that embraces the link of the main-chains, with a key put through it, above the chain.

This bridge is used for horses and carriages, and cost 25,000 dollars (5,420*l.*).*

21. It is stated in the second volume of *Histoire de la Navigation Intérieure*, published in 1820 by Mons. Cordier, that since the date of Mr. Finlay's patent forty bridges of suspension on his plan had been built. He mentions one constructed in 1815 on the Lehecgh, a mile below Northampton. It consists of two openings and two semi-arches. Its whole length is 475 feet. The chains are placed so as to divide the platform into four ways, two carriage roads in the middle, and a footpath on each side. The chains were made with iron bars 1⅜ square. The bridge cost 20,000 dollars (4,336*l.*).

22. The subject of chain bridges was, as stated in article 16., very little attended to by the English engineers until the year 1814, when a plan was brought forward by a Mr. Dumbell, of Warrington, for making a direct road from Runcorn, in Cheshire, across the Mersey to Liverpool. This scheme included a bridge in lieu of the ferry across Runcorn-gap ; and though no design had then been made, an idea was thrown out of stretching across the river a web of metallic rings.†

23. From the nature of the navigation, it was necessary that the bridge should consist of not more than three openings, the centre one of 1000 feet, and the other two of 500 feet each. And it was also necessary that there should be at least 70 feet in height under the bridge above high water.

With these requisites it appeared impracticable to construct an arch bridge. And Mr. Telford being consulted on the project, proposed an iron suspension

* Navier, p. 27 — 29.
† Provis's Account of Menai Bridge, p. 13.

bridge, of which Professor Barlow, in his Essay on the Strength of Timber, gives an account.*

24. Mr. Telford's bridge was to consist of 16 iron cables, each formed of 36 square half inch iron bars; and of the segments of cylinders proper for forming them into one immense cylindrical iron cable, which was to be nearly half a mile long (including the fixings on shore), and about $4\frac{1}{4}$ inches diameter.

The half inch bars, as well as the four segments, were to be welded together into so many lines of bars, and laid in a bundle together, secured by bucklings every 5 feet; and wrapped in flannel, well saturated with a composition of rosin and bees wax, to preserve them from the weather. They were further to be bound together with wire of about $\frac{1}{10}$ inch diameter.

The roadways were to be suspended from 16 of these cables, and to consist of two carriage-ways, and one central footpath. The main suspension piers of the middle opening were to be about 140 feet high; and the deflection of the cables in the middle $\frac{1}{20}$th of the opening, or 50 feet.

The two side openings were to consist of two semi-catenaries.

25. At the time this bridge was proposed, there existed very little experience on the construction of suspension bridges; and in order to obtain data for proportioning the strength of the parts, Mr. Telford undertook a course of experiments, of which some of the results will be stated in the next section.

* Barlow on the Strength and Stress of Timber. Third Edition. 1826. Appendix.

SECTION II.

EXPERIMENTS ON THE STRENGTH OF IRON WIRES, IRON BARS AND CHAINS, AND STEEL BARS.

26. A PART of Mr. Telford's experiments was made to ascertain the direct strength of cohesion of malleable iron ; and others were made upon iron wires stretched between two props, and loaded with weights between the props (from 0 to 258 lbs.) according to the strength of the wire. In some of the experiments both ends of the wire were fixed ; in others one end only was fixed ; and to the other was suspended a balance weight on the opposite side of the prop over which that end passed. They were made with various spans, viz. 100 feet, 31½ feet, 140 feet, and 900 feet. And also with various deflections from ⅛th of an inch in 31½ feet, upwards, according to the length of the wires, and the deflecting and balance weights.*

27. The following table contains the results of the experiments on the direct strength of cohesion of iron wire.

* Mr. Barlow has given tables of all these experiments, in his work, to which the author refers those who wish to examine them minutely.

No. of Experiments.	Section of Wire in square inches.	Weight that broke it in lbs.	Breaking strain, calculated per square inch, in tons.
I.	·00672	531 =	35·7
II.	·007854	738 =	42·0
III.	·00288	277 =	42·9
IV.	·0018	157 =	38·1
V.	·007854	630 =	35·8
VI.	·007854	634 =	36·1
			6)230·6
			Mean 38·4

28. Whence the mean vertical strength of iron wire, not more than $\frac{1}{10}$ inch diameter, may be taken at 38½ tons per square inch.

29. Colonel Dufour of Geneva made a set of experiments on the direct strength of cohesion of iron wire, of which the following is a table in English measures.*

Diameter in inches.	Section in sq. inches.	Ultimate strength in lbs. avoir.	Strength in tons, per square inch.
·03346	·000879	96	48¾
·0748	·00439	413	42
·0826	·00537	462·89	38·4
·108	·00915	808·86	39
·146	·0167	1502	40¼

It is observable that, in both these sets of experiments, the small wires appear to possess a greater strength per square inch than the large ones.

Mons. Vicat, a French engineer, found the ultimate strength of iron wire ·1181 in. diameter, 1165 lbs.,

---

* See the original in Description du Pont en Fil de Fer à Genève, par le Colonel Dufour.

which is at the rate of 47 tons per square inch; but this is so much above the results of the other experiments for wires above $\frac{1}{10}$ inch diameter, that it cannot safely be taken as a general standard. 38½ tons per square inch may be used safely as the measure of ultimate strength in computing the strength of small wire, but it is very much above the strength of a wrought iron bar containing a square inch.

30. *Experiments on the strength of iron bars.*— Mr. Telford made some experiments upon the direct strength of cohesion of malleable iron, at Messrs. Brunton and Co.'s chain cable manufactory.*

Nine experiments were made upon iron of different qualities, and of different sizes. The mean result was, that the ultimate strength of a bar of good malleable iron one inch square, is 29 tons 5⅔ cwt.

31. This is, however, presumed to be too high a result, because the machine used to try the strength of the iron was an hydraulic press, in which, with great pressures, there is great friction of the piston in the barrel of the pump. Hence the power applied to the machine, has to overcome more than the actual resistance of the iron bar that is proved by it; and, consequently, the power indicated by the machine, as being exerted to tear the iron bar asunder, is greater than the real strength of the bar.

32. A set of experiments was also made†, at Captain Brown's patent iron cable manufactory, on the strength of iron bars and cables. ‡

---

* For an account of these experiments, see Barlow, p. 261.

† See Barlow, pp. 265—269. for an account of them.

‡ The proving machine, used by Capt. Brown, is very simple and effective. It consists of two rollers set up in bearings, one at each end of a long iron frame. On the axis of one of the rollers a strong lever is fixed fast, projecting out horizontally;

c

33. The ultimate strength of a bar of good malleable iron, deduced from Captain Brown's experiments, is 25 tons. And the ultimate strength of iron bent into the form of links for chain cables, is 76 tons for a double bolt of $1\frac{1}{2}$ inch diameter (in the cable form, without stays). This is 76 tons for $3\frac{1}{2}$ square inches, or at the rate of $21\frac{1}{2}$ tons per square inch; whence the strength of iron is impaired about $\frac{1}{6}$th by being bent into links.

34. The mean of Mr. Telford's and Captain Brown's results is 27 tons per square inch, which is now taken as the standard for the ultimate, or breaking strength of cohesion of good malleable iron.*

---

its end is connected by a short vertical link to the short end of a counter horizontal lever, mounted on a fixed centre pin ; the long end of this lever is loaded; the proportions of the levers being such, that a very moderate weight on the end of the counter lever will exert great power to turn the roller round. A short vertical arm projects out below the circumference of the roller, to which a piece of chain is linked. To the end of that chain, one end of the bar that is to be proved is fastened, and the other end of the bar is linked to another chain, fastened to the other roller, at the opposite end of the machine. This second roller is turned round by 2, 4, or 6 men, according to the size of the bar under proof, by means of a train of 3 wheels and 3 pinions, until the loaded roller is pulled or turned round enough to raise up its balance weight, which is the measure of the strain the bar is intended to undergo. In the machine at Capt. Brown's manufactory, Mill Wall, near London, the proportions of the two levers are such as to make the effect of the balance weight 224 times its weight. And the tram of wheel-work is so proportioned, that 2 men at the handles can exert a force of 30 tons on the bar under proof.

* In iron bars, the strength per square inch varies somewhat according to size, and it has been observed, that when a bar is stretched so as to diminish its sectional area, the diminished bar will bear more, per *square inch*, than the original bar did. This corresponds perfectly with what was said of wires, of different

35. *Stretching strain of malleable iron bars.*— In the experiments made by Mr. Telford it was found that the iron began to stretch at about $\frac{3}{4}$ths of the breaking strain, and thence to $\frac{45}{100}$ths; the latter being the least, and the former the greatest stretching strain shown.

Mr. Donkin (in his evidence to the Committee of the House of Commons on the Holyhead road) says, that in experiments at which he was present, iron bars one inch square began to stretch with 16 tons. That is $\frac{14}{24}$ths, or nearly $\frac{6}{10}$ths of their ultimate strength.

36. It is said that iron has been found to stretch at 10 tons per square inch; the amount of the stretching being equal to about $\frac{1}{1630}$th part of the length of the bar. And that it will bear 9 tons per square inch without stretching at all.

Nine tons per square inch are accordingly now taken as the strain that the chains of a suspension bridge may bear continually without any injury, and it is prudent to proportion them for that strain. But we may, at the same time, conclude from the experiments first cited, and from the almost imperceptible amount of the stretching at 10 tons per square inch, that iron bars will bear at least 12 or 14 tons per square inch for a time, without being injured, although they ought not to be loaded with that strain continually.

37. *Strength of hammered iron.*— Some experiments have been made by Mr. Brunel on the strength of hammered iron. The mean of two sets of trials

---

diameters, in art. 28.; but the difference of strength is not important within the limits of the sizes usually given to iron bars for suspension bridges; and 27 tons per square inch is a safe standard.

of iron denoted *best,* reduced by hammer to $\frac{3}{8}$ths square in the middle, is 30·6 tons per square inch; and the mean of another set on iron denoted *best best,* reduced by hammer to $\frac{1}{2}$ square in the middle, is 32·3 tons per square inch.

These are, as might be expected, higher results than those of the first-mentioned experiments, it being well known that hammered iron is stronger than rolled iron.[*]

Strength of steel. — According to Mr. George Rennie's experiments, the ultimate strength of cohesion of cast steel is 134,256 lbs. per square inch.

Numerous experiments were made a few years ago on the strength of bars of steel and iron, by a German engineer, preparatory to building a suspension bridge of steel bars at Vienna.

The following are some of the experiments: —

No. I.

Name of Metal.	Least breaking strain in lbs. avoirdupois.	Greatest breaking strain in lbs.
Damascus steel - -	63,055	77,503
Cast steel - -	85,185	112,725
Raw steel - -	109,690	113,962
Brescian steel - -	73,758	101,354
Rolled sheet iron, cut lengthwise -	mean 31,128	
Ditto, crosswise -	—— 41,000	

No. II.

Name of Metal.	Breaking strain in lbs. avoirdupois.
Iron twice hammered - -	48,395
Damascus steel, once refined	80,000
Ditto, twice refined - - -	101,490

[*] For a table of Mr. Brunel's experiments, see Barlow, Appendix.

38. *Transverse strength of malleable iron.* — Some experiments were made at the Arsenal at Woolwich, by General Millar, on the transverse strength of malleable iron.

A bar of Swedish iron, 3 feet long and 1 inch square, supported at both ends, and loaded in the middle, was deflected in the middle $\frac{1}{4}$ inch by 560 lbs. ; and $\frac{1}{2}$ inch by 884 lbs. ; when relieved from that weight, it recovered its form. The mean result of the experiments (as stated by Mr. Barlow, p. 279.) is, *that a bar of wrought iron, 3 feet long and 1 inch square, has its elasticity destroyed* (that is, will be bent permanently) *by* 1000 *lbs.*

39. *Experiments on the deflecting weights that an iron wire stretched between two props will bear.**

The results of Mr. Telford's experiments on this vary exceedingly, and do not appear to follow any traceable law, as will be seen by the following selections from them.

A wire $\frac{6}{70}$ inches in diameter, whose vertical strength was 531 lbs., stretched between two props 100 feet apart, bore a deflecting weight of 116 lbs. (viz. 56 in the middle, and 30 at $\frac{1}{4}$ of the length from each prop) for a short time, and then broke. The middle deflection being 5 feet 8$\frac{3}{4}$ inches, or $\frac{1}{17\cdot4}$th of the span.

A wire $\frac{6}{70}$ in diameter, 531 lbs. vertical strength, stretched between props 31·5 feet apart, and with a middle deflection of 1 foot 10·75 inches (or $\frac{1}{16\cdot5}$ of the span), broke with a weight of 130$\frac{1}{2}$ lbs. suspended from the middle of the wire.

* For tables of them, see Barlow, p. 254.

A wire of $\frac{1}{10}$ inch diameter, vertical strength 630 lbs., stretched between props 900 feet apart, with a middle deflection of 17·34 feet (or $\frac{1}{52}$dth part of the span), broke with 73 lbs. (viz. 17 lbs. in the middle, and 28 at $\frac{1}{4}$ the length from each prop.)

40. Thus we have the following table of the strengths of iron wires to bear suspended weights.

Deflection in the middle, in parts of the chord line.	Ultimate strength of suspension, in parts of the ultimate vertical strength.
1 in 17·4	$\frac{1}{4\cdot58}$th
1 in 16·6	$\frac{1}{4\cdot67}$th
1 in 52	$\frac{1}{8\cdot63}$th

41. Mr. Telford made also experiments on the *momentum* that wires stretched will bear before breaking. It was found that a weight of 5 lbs. being attached, by a cord, to a wire whose vertical strength was 277 lbs., and being suddenly let fall through 10½ feet (giving a momentum of 220 lbs.) broke the wire. And a weight of 25 lbs. falling through 7¾ feet (momentum 556½ lbs.) broke a wire whose vertical strength was 630 lbs.*

42. The project for the Runcorn Bridge was not put in execution at the time of its first proposal, probably from dread of the great expense of the under-

---

* The momentum or energy of a body falling through the atmosphere is the *mass* or *weight, multiplied by the square root of the height it has fallen through,* × 8·021.

Example: suppose a weight of 10 tons to be raised 9 feet, and to drop thence suddenly on a bridge. The momentum is 10 × (3 × 8·021)=240·6 tons.

That is, a weight of 10 tons, so falling, would exert as great a strain to break down the bridge, as the pressure of 240·6 tons of dead weight.

taking, and it has not since been resumed. But it served to call the attention of the public and of scientific men to the subject, and may in so far be regarded as the origin of the great extension that has since been given to the suspension mode of bridge building.

SECTION III.

SMALL WIRE BRIDGES IN GREAT BRITAIN; AND INTRO-
DUCTION OF BAR CHAINS FOR SUSPENSION BRIDGES.

43. BETWEEN the date of the proposals for Runcorn Bridge and the year 1820, several small suspension bridges were erected in Great Britain.

The first was built in 1816, by Mr. Lees, a manu-
facturer at Galashiel, across Gala Water, to establish
a communication between different parts of his manufactory. It was made of slender wires ; the span was
111 feet, and the expense about 40*l.*

Another bridge of iron wires was built at King's
Meadows, across the Tweed. It is 110 feet long,
and 4 feet wide. It was constructed in 1817, by
Messrs. Redpath and Brown of Edinburgh, and cost
about 160*l.*

It is supported by two hollow cast iron columns
9 feet high, 8 inches diameter, and ¾ of an inch thick in
the metal, set up on the opposite banks of the river,
(see fig. 5.) 4 feet apart ; in each column is placed a
wrought iron bar 10 feet high, 2½ inches square, and
the suspending wires are attached each separately to
these bars by screw bolts 1 inch diameter. The
lower ends of the cast iron columns stand on a framed
wood foundation (see fig. 6.). The roadway is formed
of wrought iron frames, to which deal planks 1½ inch
thick are bolted. The side railing is made of wrought
iron rods ; the suspending wires are about $\frac{1}{10}$ of an
inch diameter, and are not arranged (as will be seen

by the figure) in a catenary, the method now universally adopted, but one end of each is fastened to
its supporting bar, and the other end is fastened to a
part of the roadway, so that the wires are a set of inclined lines, forming different angles with the vertical
supporting bars. When the wires are drawn tight
by the screw bolts, (as in figure 5.) the roadway is said
to have very little motion. The strength of this
bridge was tried very shortly after its erection, by
filling it entirely with a crowd of people, and it received no injury.*

There is another similar bridge, about 125 feet
span, at Thirlstane Castle, on the Ettrick, built by
Captain Napier, to replace a rope bridge.

44. A suspension bridge on this principle (viz.
with inclined chains) had been proposed (according to
Navier) by Mons. Poyet, a French engineer, more
than thirty years ago (see fig. 7.). The roadway was
to be supported by inclined wrought iron ties, fastened to vertical columns or masts, and radiating from
those columns to equidistant parts of the roadway.
This is, in fact, merely inverting the system of supporting the masts of a ship.

Monsieur Poyet says, in a Report published in
1821, that the openings may be carried as far as 40 or
50 metres, = 131 to 164 feet. And he proposes the
height of the masts to be about 85 feet for an opening of 65 feet.

45. The system of inclined wires had been adopted
in a bridge built at Dryburgh Abbey, in 1817,
across the Tweed, by Messrs. John and William
Smith, architects. It was 260 feet span, and 4 feet wide.

It was observed, that the bridge had a very sen-

* See Stevenson's Account, cited *antè*, p. 9.

sible vibration when crossed by any person; and the motion of the chains appeared to be easily accelerated. In 1818, six months after the completion of the bridge, a violent gale of wind caused so great a vibration of the chains, that the longest chains broke, the platform was carried away, and the whole bridge destroyed. The vertical motion of the roadway was said to be, just before breaking, as great as the horizontal motion, and sufficiently violent to have thrown a person off the bridge.*

46. The defect of this plan of supporting a suspended roadway by inclined chains fastened at the ends to the platform is, that the chains being of different lengths and inclinations, form different curves, and are exposed to very different degrees of tension; the long ones being the most strained, and the shortest the least. In a bridge of large span, and made with chains of any considerable weight, it would be almost impracticable to strain the inclined chains (especially the long ones) quite, or even nearly tight; for if that were done, the tension to which they would be exposed would so far absorb their strength as to leave them little power of carrying any load beyond their own weight. On the other hand, by leaving them slack enough not to impair their strength, the difference of curvature and tension of the several chains, and their want of connection and uniformity of action, cause the vibration to which they are exposed from gusts of wind, or the motion of carriages, to act unequally on the several chains; and the vibration is likely to be violent in proportion to the slackness of the chains. These reasons have probably led to the abandonment of this system for large suspension bridges.

* *Vide* Stephenson.

It appears to have answered very well for the small bridges spoken of in article 43. But the truth is, that for suspension bridges of such slight proportions, any plan almost, however defective in principle, will answer if well executed; because the strain on the bridge never bears any proportion to its strength; and, moreover, the mass of the materials suspended is generally too little to do injury by being put in motion : but, on the contrary, in suspension bridges of large dimensions, and consequently of great weight, the force that the suspended mass will acquire by being put in motion, increases rapidly. Hence it is an object to make it resist motion, and especially to make every part bear its fair share of strain. It is a common doctrine that lightness is the peculiar excellence of a suspension bridge, but that is a principle which must be acted upon with discretion, and not taken generally. For a bridge may be from its size just so heavy that by being put in motion it will acquire great momentum, and just so light and slight, that it will be unable to resist the effects of its own vibration. Therefore, when it becomes necessary to make the chains of a bridge so heavy that vibration would be dangerous, it is advisable boldly to increase their weight, rather than attempt to diminish it, and to bind and connect the several chains and the road way firmly together, in order that there may be sufficient *mass and stiffness* in the bridge to *resist* motion, rather than yield to it readily.

47. " The effect we have to provide against on " bridges of suspension (says Mr. Stevenson) is not " merely what is technically termed *dead weight*. A " more powerful agent exists in the sudden impulses " or jerking motion of the load, which we have partly " noticed in the description of the Dryburgh Bridge.

" The greatest trial, for example, which the timber
" bridge at Montrose, about 500 feet in extent, has
" been considered to withstand, is the passing of a
" regiment of foot, marching in regular time."

" A troop of cavalry, on the contrary, does not
" produce corresponding effects, owing to the irre-
" gular step of the horses. The same observation will
" apply to a crowd of persons walking promiscuously,
" or a drove of cattle, which counteracts the undu-
" lating and rocking motion observed on some occa-
" sions at the bridge of Montrose, when infantry
" has been passing along it. Hence, also, the effects
" of gusts of wind often and violently repeated, which
" destroy the equilibrium of the parts of a bridge of
" suspension; and the importance of having the
" whole roadway and side-rails formed in the strongest
" possible manner."

48. In the Dryburgh Bridge the links of the
chains were formed at one end into an eye by welding.
The other end of the bar was turned back upon itself,
and held by a clamp or band of iron. In the broken
chain, only one or two links had given way at the
welded joint; all the others had broken at the other
or loose eye.

The bridge had cost in the beginning 500*l.* It
was rebuilt for an additional 220*l.*, on the catenarian
system, with bar chains. The roadway was stiffened
by a side parapet. Before this parapet was put up,
during the repairs of the bridge, a gust of wind struck
one end of the platform, and raised it up above the
level of the road.

An undulating motion was thereby produced, running all along the bridge, and striking the other with
a jerking motion; but since the fixing of the parapets,
this vibration is diminished. To steady the bridge

further, oblique ties are attached to the bridge, and
to piles on the bank. This is said to answer its intended purpose, and to diminish the motion of the
bridge in high winds.*

49. The new Dryburgh Bridge (see fig. 8.) is
supported by four main chains, two side by side over
each suspension pillar. The chains are made of
round iron rods, about $1\frac{5}{8}$ inches diameter, and about
10 feet long. The eyes at their ends are united by
oval coupling links, about 9 inches long. The roadway is suspended by vertical rods $\frac{1}{2}$ inch diameter,
the upper ends resting on the coupling links with a
sort of cross-head, and the lower ends passing through
the side bearers of the roadway, and being screwed
up under them with nuts.

The pillars of suspension are Memel piles 9 feet
apart. They are united by a cross beam at the top,
over which the chains pass. The chains are 12 feet
apart at the approaches; but only $4\frac{1}{2}$ at the middle of
the bridge, viz. the width of the roadway, which is $4\frac{1}{2}$
feet wide throughout, so that the suspending rods are
not quite vertical. The object of this arrangement
is to stiffen the bridge, but it must tend to bring a
twisting strain on the points of suspension, and to
keep the bridge in an unnatural position. The
points of suspension are 28 feet above the level of the
road; the deflection between $\frac{1}{10}$th and $\frac{1}{11}$th of the span.
The road is about 18 feet above low water; and is
formed of two beams of deal running the whole length
of the bridge, and framed together. Cross joists are
laid on this frame. The side beams, or bearers, are
tied to the abutments by two chains of 1 inch diameter
from their undersides. The back stays are chains
of 1 inch diameter, and go down into the ground

* See Stevenson's Account.

through a mass of masonry under which they are fastened.

50. In 1820, Mr. Stevenson of Edinburgh published, in his Account of Suspension Bridges, cited *antè*, a design for a bridge, intended originally to cross the river Almond, between Edinburgh and Queensferry. Its chief peculiarity was the mode of attaching the ends of the main chains, by means of which he dispensed with the use of high suspension towers. He proposed to pass the chains round abutments or masses of masonry, built on the banks, the chains lying in a tunnel running round three sides of each abutment, to which a cross tunnel or gallery was to be built to give access to them. The chains were fastened at the face or river side of each abutment (see fig. 9.), and the ends of the chains were to be made for this purpose with conical heads, and were to pass through a conical cast iron pipe laid in the masonry.

The roadway was to be laid on cast iron frames fixed up on the main chains. The span proposed was 150 feet. Mr. Stevenson, in his paper, attributes some advantages to this construction, but deems it unadvisable for a span beyond 200 feet.*

51. This plan appears certainly unadvisable for a bridge of large dimensions. There may be situations where, in the erection of a small bridge, the expense of towers or suspension piers may be saved by it, and it may be more convenient; but for a large bridge, the great length of the cast iron standards for the

---

* A similar plan has been proposed lately by Mr. Armstrong, for a bridge of suspension at Clifton. And part of the roadway on the land side of the piers in the Hammersmith Bridge, is supported by cast iron frames, standing on the main chains. Also a small bridge has been erected on that plan in Scotland. (See *post*, Description of the Bridge at Doune.)

central part of the opening, and the strength that they would require, to be sufficiently stiff, would increase both the weight and the expense, without any corresponding advantage. There would be, besides, great difficulty in giving the frames for the road lateral stability, because they have no breadth of bearing to stand on, and their foundation is in itself unsteady, and liable to vibration. Hence any excess of weight on one side of the bridge, would act at the end of a long lever to pull over the roadway and the frames that supported it ; and it is probable that any great weight would really do so, unless firmness and stiffness were obtained by an injurious profusion of materials. In fact, a bridge of this construction must be made a great deal too heavy, in order to counteract its natural defects.

52. Wire suspension bridges are common on the Continent; but as they are mostly on the catenarian plan, and have followed in order of dates the construction of chain bridges in this country, the account of them will be deferred to another section.

53. *Introduction of Bar Chains.*

The plan proposed for the first large bridge in this country (viz. the Runcorn Bridge) was, as we have seen in art. 24., to suspend the platform from iron cables made of small bars welded together at their ends, and bound up together, so as to form a long flexible bar or cable. That plan has never come into operation, and the system universally adopted in practice, is to make the chains of straight wrought iron rods or bars, from 5 feet to 15 feet long, with either welded eyes, or holes drilled out at their ends, by which they are connected together in pairs by short links and bolt pins.

This system of making the main chains was intro-
duced by Captain Samuel Brown, R. N., who constructed a model of a suspension bridge made thus of
straight bars in 1813.   It was 105 feet span, and the
whole of the iron work weighed 37 cwt.   It bore
loaded carts and other carriages.*

54. Captain Brown took a patent in 1817 for his
invention.†   In his specification he states his object
to be to substitute for the use of common link chains
or wires, (in constructing suspension bridges,) straight
bars, round or flat, having their ends united by
coupling plates and bolt pins, or by welding (or by
some other methods which he describes), so that a
number of these straight bars united will form one
long chain.

The reasons given by Captain Brown for preferring
this mode of making the main chains of a suspension
bridge, to making them of links like chain cables, or
of small wires, is this : That link chains are weaker in
proportion to their weight than bars, a portion of the
strength of the iron being lost in bending it into the
form of a link (see *ante*, experiments on link chains,
art. 33.) ; and small wires are inexpedient for large
bridges, on account of the number of parts and complication of joinings required, and also on account of
the great surface they expose to damp, and their consequent liability to destruction.

---

* Mr. Rennie, in his evidence before the Committee of the
House of Commons, on the Menai Bridge, April, 1819, speaks
of this model, erected at Capt. Brown's manufactory in London, and says that he had a carriage driven over it several
times, and perceived very little vibration.

† Capt. Brown's specification is enrolled at the Petty Bag Of-
fice, Chancery Lane.   Patent dated 10th July, 1817.   Specification enrolled 9th January, 1818.

55. Captain Brown gives, in the specification of his patent, a very complete design for a suspension bridge with chains made of straight bars.

The span 1000 feet; the greatest deflection of the chains 40 feet, or $\frac{1}{25}$th of the chord line; the platform 30 feet wide, with a carriage way at each side 12 feet broad, and a central footpath 6 feet broad.

There were to be eight double lines of chains, or sixteen chains in all, each consisting of a series of bars about 14 feet long and 6 inches broad by 2 inches thick. The total section of the main chains, therefore, would be $(16 \times 12 =)$ 192 square inches.

The bars to be of the form shown in fig.10. plate 1., with holes punched out at their ends, leaving rather more metal round the ends than in the middle of the bar, and to be connected by coupling plates and bolt pins, shown also in fig. 10. The coupling plates and bars to be bound round by a hoop over each end of the coupling plate. (See the lower figure in fig. 10., which represents two of the bars connected.)

The platform to be laid on longitudinal chains, or iron bearers, suspended by vertical rods from the main chains. These longitudinal bearers to be formed of bars about 7 feet long, coupled like the main chains, but only of proportionate strength to bear the weight of the roadway. On them to be laid cross joists of wood in pairs, each joist about 12 inches deep by 3 inches wide, leaving a little space between them, through which the vertical suspending rods descend to lay hold of the longitudinal roadway bearers. On the cross joists a course of 3 inch planking to be laid, spiked down to them, and iron guards to be fastened to it at the sides of the carriage road.

The platform to rise upwards in a curve, which Captain Brown directs to be at the rate of about 25 feet in the middle for 1000 feet of span.

The supension piers to be massive arches of stone ; the back stays are carried over them, and thence to the abutments, in cast-iron pipes laid in the masonry ; and are thence directed to be carried down some depth into the ground through a solid mass of masonry, if the shores do not offer solid rock for the abutments. The ends of the chains to be held by strong fastening bolts put through them outside of large cast-iron plates, bearing against the masonry of the abutments.

Captain Brown states, in his specification, that he had made experiments on which he had founded designs and calculations for suspension bridges exactly on the above plan as early as 1808, and that he made the model mentioned *ante*, art. 53. in 1813.

56. It is, indeed, sufficiently evident from the tenor of the directions given in his specification for constructing a suspension bridge, that they must have been the result of much thought and experiment ; for if his design be compared with the bridges since erected (of which descriptions will be given in the subsequent parts of this work), it will be found that they are all made upon the same system of putting the chains together, and do not offer any striking difference in proportions, except in the deflection, which is now hardly ever made less than $\frac{1}{15}$th of the chord line.

57. Figures 11, 12, 13. and 15. are sketches of the several plans proposed by Captain Brown in his specification for coupling the links of his bar chains.

In figure 11. the chains are proposed to be made of round rods ; each end of the rod is swelled out into a cone, and is received in a cast iron socket made in two halves. (See fig. 11.) When the two halves of the conical socket are put together over the conical ends of the rods, they are bound round with hoops, and make thus a firm joint. Captain Brown states this to be the method of fastening that he had adopted for proving bars by a longitudinal strain, in a proving machine, and that they never gave way at the joints.

In fig. 12. the chains are proposed to be made of round iron, with welded eyes at the ends ; and the coupling links are also of round iron welded into the shape of a link, to receive cross pins.

In fig. 13., 16 small bars, each $\frac{7}{8}$ inch square, are proposed to be laid together to form a chain. Every bar to have a jagged scarf, or teeth upon it at some part of its length, by which means the bars catch against each other, and are prevented from drawing away endwise. And the whole system of bars to be bound round by hoops at proper distances, taking care that the jagged scarfs be arranged at break joint, so that there shall never be more than one joining between the same pair of hoops.

Fig. 15. shows a plan for taking out a damaged link from a chain.

A pair of half links *a a*, long enough to enclose three entire links *c e d*, are placed outside of the chain. The ends of the half links *a a*, catching against the ends of the coupling plates *i*, and being held together by a hoop jammed on at each end ; they form thus a long link, connecting the two outer links, without depending upon the intermediate three links *c e d*; and the middle one (*e*) can be taken out to replace or repair it, by withdrawing its bolt pins through the holes at *h h* in the half link (*a a*).

58. Of these various plans only two have been adopted in practice by Captain Brown, and by other engineers, viz. those shown in figures 10. and 12., which are the most convenient for making and fixing the chains ; the least liable to be deranged by strains or shocks ; and the best adapted for taking out and repairing a link when damaged.

The plan (fig. 11.) may answer very well for a chain that is to be kept tight, as it would be in proving a bar ; but it is doubtful whether it would admit of sufficient flexibility in the chain for a suspension bridge, unless the couplings were made with a good deal of play, and then they would hardly be secure.

DESCRIPTION OF THE UNION SUSPENSION BRIDGE OVER
THE TWEED.

(Plate II. Figures 1. to 6.)

59. The Union Bridge across the Tweed, five miles above Berwick, designed and executed by Captain S. Brown, R. N., was the first large bar chain bridge completed in this country.

It was begun in August, 1819, and opened in July, 1820, just after the commencement of the Menai Bridge.

The following description of it is taken from Mr. Stevenson's Account, published in the Edinburgh Philosophical Journal, No. X., and from the account given by Monsieur Navier, who examined it in 1821. The chord line, or distance between the points of suspension, is 449 feet; and the deflection about 30 feet.

The main chains are 12 in number, arranged in pairs, and placed in 3 ranges, one under the other, on each side of the bridge, about 1 foot 7 inches apart. Each link of the chains is a round rod, 2 inches diameter, and 15 feet long, of the best Welsh iron, formed into an eye at each end by welding. The long links are connected together by short open coupling links 6¾ long centre and centre, made of iron 1⅛ square, and united to the eyes of the long links by oval bolt pins 2½ by 2 inches, which pass through both; (see figs. 4. and 5., which are a plan and side view of one of the couplings.) The bolt pins have a head at one end, and are keyed at the other. The coupling joints of the main chains which support the vertical suspending rods, are arranged so that the vertical sus-

pending rods are fastened alternately to each row
of chains ; viz. the first to the lowest chain ; the
second to the chain next above ; and the third to the
upper chain. Hence the weight of the bridge is
borne equally by all the chains, or is distributed
equally over their whole length.

The main suspension pier on the Scotch side is a
pillar of Aisler masonry. It is slightly pyramidal,
and is 60 feet high, about 36 feet broad, and 17½
feet thick at its medium dimensions, according to
Mr. Stevenson. The arched opening through it is
12 feet wide, and 17 feet high.

On the English side the suspension pier is built in
an excavation of a rock of precipitous sandstone.
Its height is about 20 feet ; and it is of the same
shape as the upper part of the opposite pier.

The chains pass through openings in the main pier
on the Scotch side, one above the other, about 2 feet
apart, and rest therein upon rollers laid in the masonry. The links of the chains are made very short
at that part, that they may lie on the rollers without
being bent.

From the points of suspension the chains pass down
24 feet into the ground, and through large iron holding plates, to which they are fastened by a strong
oval bolt 3 inches by 3½.

The holding plates are 6 feet by 5 feet ; 5 inches
thick in the middle, and 2½ at the edges. They
are kept down by large mound stones, and other
materials piled up to the level of the road.

On the English side the chains lie also upon cast
iron plates laid in the masonry. The extremities of
the chains are fastened, as on the other side, to large
holding plates of the same dimensions ; but instead
of being sunk deep in the ground, they are placed

vertically rather above the level of the foundation of
the pier, and they bear against a horizontal arch of
masonry dovetailed into the rock.

60. The roadway (see fig. 6., which is a section of
the platform,) is of deal; that part which is for the
carriage road being covered with iron tracks.

It is 387 feet long (see figs. 1. and 2.), and 18
feet wide between the parapets.

The carriage-way in the middle is 12 feet wide,
and the footpath on each side 3 feet wide.

The carriage-way is protected by cast iron guards,
4¼ inches high, and is covered by iron straps, about
¾ths of an inch thick, placed lengthwise on like a railroad, in the track of the wheels, and crosswise in the
track of the horses' feet.

The platform is borne by 2 iron side bars, or
bearers, 3 inches deep by ⅞ths thick, extending all
the length of the bridge, and supported by the vertical suspension rods.

On the side bearers are laid cross joists, 15 inches
deep by 7 inches; and over these a course of 3 inch
deal planking is laid, lengthwise, for the roadway.

The platform is curved a little upwards (viz.
about 2 feet in the centre) from the horizontal line.

The vertical suspending rods are round iron, 1
inch in diameter. At the upper ends they spread
out into a dovetail, which is wedged in a hole in a
cast iron cap, or saddle, so that the suspending rod
cannot draw out of it. (See fig. 5.)

The cast iron caps are of the form shown by
figures 4. and 5. They rest upon the chains at the
couplings, bearing on the ends of each pair of long
links, and on the coupling bolts. The ends of the
caps overlap these bolts a little, and the middle

part of the cap descends between the ends of the long links, and keeps them apart.

The lower ends of the suspending rods are a little larger than the upper ends, and spread out into the shape of a fork, which clasps the iron side bar, and they are held by keys and keepers underneath. (See fig. 6.)

The vertical suspending rods are 5 feet apart.

The 12 main chains, with the parts belonging to them, weigh about 5 tons each; and the whole weight suspended is estimated at 100 tons.

The parapets are 5 feet high; they are formed by several rows of horizontal rods, 1 inch diameter, which connect the vertical suspending rods together, and are fastened also to other vertical standards, placed one between each pair of vertical suspending rods.

The upper and lower rows of horizontal rods are flat iron, $1\frac{3}{4}$ by $\frac{1}{2}$ thick; the suspending rods pass through holes in the horizontal tie rods, and the latter pass through openings in the vertical standards between the suspending rods. Hence the suspending rods are tied to the parapet.

61. The Union Bridge was exposed to a severe trial at its very opening, but was not at all injured by it. The crowd of spectators broke through the toll-gates, and filled the bridge to the number, it is stated, of 700 people. Reckoning each person at 150 lbs. weight, the whole was 47 tons: add 100 tons (the weight of the bridge), the whole weight supported by the chains was 147 tons.

The deflection of the chains is $\frac{449}{30} = \frac{1}{14\cdot96}$th, and the tension at the points of suspension for that deflection is $1\cdot923$ times the weight suspended, or $147 \times 1\cdot923 = 282\frac{1}{2}$ tons.

The chains contain 37·68 square inches of iron, of which the ultimate strength is $37·68 \times 27 = 1017$ tons; and taking 9 tons per square inch as the load that iron will bear without any stretching, we have $37·68 \times 9 = 339·12$ tons for the strain that the Union Bridge may bear constantly without injury.

Mr. Stevenson calculates its strength by an experiment made at Messrs. Brunton's cable manufactory, London, in which the ultimate strength of a bar of iron about 2 inches diameter was found to be 92 tons; whence the ultimate strength of the 12 bars of the Union Bridge is $(12 \times 92) = 1104$ tons.

DESCRIPTION OF THE TRINITY SUSPENSION PIER AT NEW-
HAVEN, IN THE FRITH OF FORTH, BUILT BY CAPTAIN
S. BROWN, R.N.

(Figures 7. to 12. Plate II.)

62. The Newhaven suspension pier was undertaken in 1821, at the expense of the proprietors of
the steam vessels employed in the Frith of Forth,
and of several gentlemen forming the Trinity Pier
Company.

The great increase of intercourse with the Scotch
coast, brought about by the introduction of steam
boats within a few years after the close of the
continental war, had made it an object with the
proprietors to increase the facility of landing and
embarking; and not being able to enter into a
satisfactory arrangement with the proprietors of the
Newhaven stone pier, it was proposed by Lieutenant
Crichton, R. N., one of the chief agents of the
London and Edinburgh Steam Navigation Company,
to erect a new pier. This was decided upon, and
the execution was intrusted, by the speculators, to
Captain Brown, who began to drive piles for the
main piers in March 1821, and completed that part
of the work in July of the same year. The following
account of this work is taken from a paper, by Captain Brown, printed in the Edinburgh Philosophical
Journal, No. XI., and from Monsieur Navier's *Mémoire sur les Ponts Suspendus.*

The extreme length of the pier, from high-water
mark to the end of the pier, is 700 feet; its width
4 feet. It is in three divisions, each of 209 feet

span, and 14 feet deflection, without any intermediate
support ; height above high water 10 feet. The pier
head is 60 feet wide by 50 feet long, supported by
46 piles, driven about 8 feet into stiff blue clay.
The heads of the piles are tied together by beams at
right angles, and by diagonal trusses and warping,
which form, at the same time, a secure framing for
the deck, made of 2-inch plank.

The front of the pier faces the N. E., and is
exposed to the whole range of the sea from the
entrance of the Forth. It has also to sustain the
drag of the bridge, and therefore it is strongly
sustained, and backed by diagonal shores driven in
opposite directions.

The intermediate piers are only subject to pressure
from the weight of their respective divisions, and are
greatly sheltered from the swell by the outer pier.
Their area, therefore, is merely sufficient to form a
secure framing for the cast-iron standards, over
which the main chains or suspending bars are supported.

The land pier is a stone pillar of solid masonry,
6 feet square and 20 feet high. The main bars pass
over the top of this pier. The back stays form an
angle of 45° to the horizontal chord line. Their
extremities are sunk about 10 feet below the surface
of the ground, and are secured in hard clay by cast
iron plates, on the principle of the mushroom anchor.
The outer back stays are carried at the same angle
over the standard of the pier head, and are moored
into a rider, which is locked to the piles, and these
riders are backed by spur shores to resist the drag of
the bridge.

There are two main chains formed of eye bolts, or
round rods, about 10 feet long, 1⅛ and 1   diameter,

being of different dimensions, in order to have strong bolts at the points of suspension where the strain is greatest, and to diminish the strength of the bolts towards the centre, where the strain is least; but it was not attempted to make the links throughout exactly proportional to the strain they would have to bear. The links, or long eye-bolts, are united by coupling links and bolt-pins of proportionate strength. The road is borne, as in the Union Bridge, by two long iron side bars, or roadway bearers, which are 3 inches deep by   thick; their extremities overlap each other, and are bolted together. They are supported by vertical suspending rods which pass up between the coupling links at the joints of the main chains, and are held by a cross key (see figs. 11. and 12.); and the cross beams or joists for the road are laid upon the side bearers, and covered with 2-inch planking. On each side is a wrought iron parapet about 4 feet high; and the vertical suspending rods form the standards for the horizontal rails of the parapet. (See fig. 10.)

63. Captain Brown had found, by many experiments, that a round rod $1\frac{3}{4}$ diameter, broke with 147,000 lbs., or 65·6 tons, and that it began to stretch with $\frac{3}{8}$ths of that strain. He therefore proved each of the main chains of the Newhaven Pier with a strain of 88,200 lbs. $= 39·3$ tons; that is, $16\frac{1}{4}$ tons per square inch for the $1\frac{3}{4}$ links, and 14 tons per square inch for the $1\frac{7}{8}$ links: the main chains containing 5·52 square inches of iron at their strongest part, and 4·86 square inches at their weakest part.

" The pier," says Captain Brown, " has borne, " since its erection, a load of 21 tons, besides the " usual weight of persons passing over it, a load, " probably, greater than it will have to bear."

The cast-iron standards that carry the main chains consist of two triangular frames (see figs. 8. and 9.), each cast in one piece. The section of the ribs of the frame is a T, and they are tied together in the direction of the length of the road by diagonal braces bolted to the broad edges of the T. (See fig. 8.)

The main chains lie in cast-iron saddles, or sockets, placed on the tops of the suspension standards. (See fig. 9.) The roadway is braced by inclined ties (see fig. 7.), made of eye bolts 1 inch diameter, and about 13 feet long, connected by coupling links.

The upper ends of these ties pass through wrought iron ears bolted to the main standards, and are held to them by tightening nuts ; and the lower ends of the ties lay hold of the platform. There is also a set of inclined ties under the bridge fastened to the piles. (See fig. 7.)*

The suspension frames stand on blocks of wood, and are bolted through them to the piles.

N. B. In Monsieur Navier's drawings, from which the figures of the Newhaven Pier are taken, the points of suspension are about 16¾ feet only above the road, and the chains dip to within 3¾ feet of the road in the middle of each opening. The deflection is only 12½ feet ; the road, therefore, rises very little in the middle. These measures do not agree with those stated by Captain Brown in his own account.

The Newhaven Pier is a very light construction, and vibrates sensibly with the passage of a single person. It has, however, stood more than ten years, during which it has weathered some severe gales, without showing any signs of failure.

* The upper ties are not mentioned by Capt. Brown, but were put up before 1821. The lower ties were not put in when Mons. Navier examined the pier, in 1821, and have been put in since.

PROPOSALS FOR A BRIDGE OVER THE MENAI STRAITS.

64. The Menai Strait runs nearly S.W. and N.E. between the Island of Anglesea and the opposite coast of Carnarvon.

The connection between the two shores was kept up before the erection of the bridge by six ferries, the most important of which was that about a mile north of the Swellies, called *Bangor Ferry*, from its proximity to the city of Bangor.*

The idea of establishing a permanent roadway, instead of the ferry, was entertained as early as 1785, and various schemes were proposed for its accomplishment. None were, however, pursued seriously until after the union of Ireland with England. Then, by reason of the increased and increasing intercourse between the two countries, the inconvenience of the ferry became more felt; and in consequence Mr. Rennie was directed by government in 1801 to survey the Menai Strait, and propose a plan and estimate for a bridge.

65. Mr. Rennie prepared four designs, two for crossing at the Swellies, and two for crossing at Ynys y Moch, or Pig's Island. (See the Map, fig. 1., Plate III.)

The first design for Ynys y Moch, consisted of a central cast iron arch, 450 feet span, and 150 feet in height from high water to the under side of the arch, to communicate with the shores on either side by a series of gradually diminishing stone arches. The estimate was 259,140*l.*

---

* The *Swellies* are rocks in the Strait, extending about two miles S. W. of Bangor Ferry.

The second was a central cast iron arch the same
as the first, with two other iron side arches, each 180
feet span, to be connected also with the shores by
stone arches.   Estimate, 262,500*l.*

For crossing at the Swellies, the first design was a
central cast-iron arch, 350 feet span, with the crown
of the arch 150 feet above high water; a stone arch
on each side 100 feet span, and two other cast-iron
arches of 300 feet span at the ends of the stone ones.
Estimate, 265,812*l.*

The second, three cast-iron arches, each 350 feet
span, and 150 feet in height from high water. Estimate, 290,417*l.*

While these designs were under consideration, opposition to the erection of a bridge arose from parties
interested, and from that and pecuniary consider-
ations, the plan was suspended until 1810, when a
Parliamentary Committee was appointed to examine
into the state of the road from Shrewsbury and Chester to Holyhead.

Mr. Rennie's designs were then again brought under
consideration, and evidence was taken from nautical
men as to the effect of a bridge upon the navigation.

The evidence was, as might have been expected,
exceedingly contradictory, some saying that a bridge
would destroy the navigation, others that it would do
no harm whatever. The result, however, of the
investigation, was a report to Parliament by the
Committee, that in their opinion no injury could be
done to the navigation of the Menai Straits by the
construction of the proposed bridge at the Swellies,
and that such a bridge was expedient, and ought to
be immediately constructed.* They recommended

* See Second Report of the Committee on the Holyhead
Road and Harbour, 1810.

Mr. Rennie's design mentioned above with three cast-iron arches over the Swellies. (Figures 2. and 3. are an elevation and plan of it.)

66. In consequence of this report, the Lords of the Treasury directed Mr. Telford to survey the Holyhead road, and to consider the best mode of passing the Menai Strait. This was done, and in Mr. Telford's report to the Lords of the Treasury, (April, 1811,) he proposed a cast-iron bridge of one arch, 500 feet span, 100 feet clear in height at high spring tides, and 40 feet broad. (See fig. 4.) Estimate, 127,331*l*.

It being nearly impracticable to construct the centering of the arch from below, on account of the nature of the bottom of the channel, and the great rise and rapidity of the tides, Mr. Telford proposed to suspend the centering from above. (See fig. 5.)

Mr. Telford gives the following description of this mode of constructing the centering. (See his Report in the Appendix to the Report of the Committee on Holyhead Roads, 1811. P. 87.) *

" I propose, in the first place, to build the masonry
" of the abutments as far back as A. (See fig. 5.
Plate III.)

" Having carried up the masonry to the level of
" the roadway, I propose, upon the top of each
" abutment, to construct as many frames as there are
" to be ribs in the centres, and of at least an equal
" breadth with the top of each rib. These frames
" to be about fifty feet high above the top of the
" masonry, and to be rendered perfectly firm and

* The design for this bridge was very bold, and the plan proposed for suspending the centering was ingenious, and possibly paved the way for the adoption of the plan ultimately fixed upon of constructing a suspension bridge over the Menai Straits.

" secure. That this can be done, is so evident, I
" avoid entering into details respecting the mode.
" These frames are for the purpose of receiving
" strong blocks, or rollers and chains, and to be
" acted upon by windlasses or other powers.

" I next proceed to construct the centering itself:
" it is proposed to be made of deal baulk, and to
" consist of four separate ribs, each rib consisting of
" a continuation of timber frames, five feet in width
" across the top and bottom, and varying in depth
" from 25 feet, near the abutment, to 7 feet 6
" inches at the middle or crown. Next to the face
" of the abutment, one set of frames about 50 feet in
" length can, by means of temporary scaffolding and
" iron chain bars, be readily constructed and fixed
" upon the masonry offsets of the abutment, and to
" horizontal iron ties laid into the masonry for this
" purpose. A set of these frames (four in number)
" having been fixed against the face of each abut-
" ment, they are to be secured together by cross and
" diagonal braces; and there being spaces of only 6
" feet 8 inches left between the ribs (of which these
" frames are the commencement), they are to be co-
" vered with planking, and the whole converted into
" a platform 50 feet by 40. By the nature of the
" framing, and from its being secured by horizontal
" and suspending bars, I presume every person accus-
" tomed to practical operations will admit that these
" platforms may be rendered perfectly firm and
" secure.

" The second portion of the centering frames
" having been previously prepared and fitted together
" in the carpenter's yard, are brought in separate
" pieces, through passages purposely left open in the

E

" masonry, to the before-mentioned platform ; they
" are here put together, and each frame raised by
" the suspending chain bars and other means, so that
" the end which is to be joined to the frame already
" fixed shall rest upon a small moveable carriage : it
" is then to be pushed forward, perhaps upon an iron
" railroad, until the strong iron forks which are fixed
" upon its edge shall fall upon a round iron bar
" which forms the outer edge of the first or abutment
" frames : when this has been done, strong iron bolts
" are put through eyes in the forks, and the afore-
" said second portion of framework is suffered to
" descend to its intended position by means of the
" suspending chain bars, until it closes with the end
" of the previously fixed frame like a rule joint. Ad-
" mitting the first frames were firmly fixed, and that
" the hinge part of this joint is sufficiently strong,
" and the joint itself about 20 feet deep, I conceive
" that even without the aid of the suspending bars,
" this second portion of the centering would be
" supported ; but we will for a moment suppose that
" it is to be wholly suspended.— It is known by ex-
" periments, that a bar of good malleable iron, one
" inch square, will suspend 80,000 lbs., and that the
" powers of suspension are as the sections ; conse-
" quently a bar of 1½ inch square will suspend
" 180,000 lbs. ; but the whole weight of this portion
" of rib, including the weight of the suspending bar,
" is only about 30,000 lbs., or one-sixth of the
" weight that might be safely suspended ; and as I
" propose two suspending chain bars to each portion
" of rib, if they had the whole to support they would
" only be exerting about $\frac{1}{12}$th of their power ; and
" considering the proportion of the weight which
" rests upon the abutments, they are equal also to

" support all the ironwork of the bridge, and be
" still far within their power.

" Having thus provided for the second portions of
" the centering a degree of security far beyond what
" can be required, similar operations are carried on
" from each abutment until the parts are joined in
" the middle and form a complete centering; and
" being then braced together, and covered with
" planking where necessary, they become one general
" platform or wooden bridge on which to lay the
" ironwork."

67. The Parliamentary Committee having heard
further evidence as to the effect of a bridge over the
Menai upon the navigation, and more particularly as
to the effect of Mr. Telford's plan, made their report
in May, 1811, in which they recommended in the
strongest manner to the House the adoption of such
measures as might give to the public the benefit of
the proposed plan. *

This recommendation was not, however, then
acted upon, and the design lay dormant until 1818.

In the meanwhile the Runcorn Bridge had been
proposed by Mr. Telford, and Captain Brown had
constructed the model of a suspension bridge mentioned in art. 53. Also some small wire bridges
had been constructed.

The public attention had thus been drawn to suspension bridges; and the improvements of the Holyhead road being in active progress (in pursuance of
an act passed in 1815) under the direction of Mr.
Telford, he was ordered in 1818 by government to
give his opinion whether a suspension bridge was
practicable over the Menai Straits, and if he considered it so, to prepare a design.

68. Mr. Telford accordingly again surveyed the localities, and proposed a suspension bridge at Ynys y Moch. (See fig. 6.) The distance between the supporting pyramids to be 560 feet ; the height of the pyramids 50 feet above the roadway; the height of the roadway above the top of spring tide 100 feet. The deflection or versed sine of the arch was to be 37 feet, or $\frac{1}{15}$th of the chord line. The breadth of the bridge to be about 30 feet, having two carriage ways of 12 feet each, and a footpath of 4 feet between them. The main chains were to be 16 in number, each composed of 36 bars of half-inch square iron, and made up with segments to a cable nearly 4 inches diameter, in the way proposed for the Runcorn Bridge. * Each bar was to be welded together for the whole length of the chain, and to be further secured by bucklings. The whole was then to be bound round by small iron wire, and coated with some substance to preserve it from the action of the atmosphere.

The whole section of the chains was 192 square inches.

The weight of the cables was rated at 342 tons ; and the whole weight of the bridge, roadway and chains, was rated 489 tons ; and add 300 tons for load, that would make in all 789 tons. †

69. Mr. Telford had found by experiment that with a chord line of 560 feet, and a deflection of 37 feet, or of $\frac{1}{15}$th, a bar of good iron 1 inch square would bear besides its own weight $10\frac{1}{2}$ tons, and half that before it began to stretch. Hence, with a section of 192 square inches, this bridge would bear without

* See *antè*, art. 24.

† See Mr. Telford's Report, April 1819; Third Report of the Committee on Holyhead Roads, 1819.

injury ($192 \times 5\frac{1}{4} =$) 1008 tons, or a surplus of 519 tons above the real weight of the suspended part ; and there would have been required a further weight of 1008 tons to break it down.

This was the original outline of the design for the Menai Bridge ; which was departed from in many points, but adhered to in its general features, as we shall see further.

Evidence was taken upon it by the Committee, at considerable length, as to the practicability of Mr. Telford's design ; and the opinions of engineers being favourable, a report was made, in consequence of which 20,000*l.* were voted by Parliament to enable the Commissioners to commence operations. Mr. Telford was accordingly directed to take measures for commencing the work, which he did in July, 1818.

70. The opposition of parties interested was not however at an end ; and, consequently, although the preparatory operations were not suspended, the building of the bridge was delayed until the Committee, having furnished themselves with adequate information, should have obtained an act to confirm and enlarge their powers.

This act was brought into the House in 1819, and the evidence gone into having been such as to satisfy the Committee of the advantage of the bridge, they made their report accordingly ; and an act was at length passed, in July, 1819, giving propei powers for carrying on the erection of the bridge.

DESCRIPTION OF THE MENAI BRIDGE, DESIGNED AND
ERECTED BY THOMAS TELFORD, ESQ.: COMMENCED
AUGUST, 1819; OPENED JANUARY, 1826.

(Plate IV.)

71. The distance apart of the centres of   *Feet. Inches.*

the summits of the main pyramids is - 579 10½

The deflection of the chains in the middle   43   0

The clear height of the roadway above

   high water   -   -   -   102   0

Breadth of the platform   -   -   28   0

It is divided into two carriage-ways, one on each
side of the bridge, 12 feet wide, and a central footpath, 4 feet wide.

72. *Main piers.*—The piers of suspension are pyramids, built of grey marble from the Penmon quar-
ries, near Beaumaris, in the island of Anglesey. The
Anglesey main pier stands on the rock called *Ynys
y Moch,* (see the Map, fig. 1. Plate III.) which was
levelled by blasting to form a foundation, and rises to
about the level of high water.

On the Carnarvon side, the foundation of the pier
is sunk 7 feet below the surface of the beach, and is
also firm rock.

The main piers are 100 feet high, from the level of
high water line up to the level of the roadway, and
their summits are 53 feet above the roadway.

The shape of the piers below the roadway is octa-
gonal ; their extreme breadth at the base 70 feet, and
at the summit 45 feet. Their thickness at the base
is 50 feet, at the level of the roadway 29 feet, and at
the summit 11 feet. They are not built solid all the
way up, but in each four hollow squares, about 9½
feet square, were left, commencing above high water

mark, and running up to within about 4 feet of the level of the roadway.

Through each main pier two arched openings are formed for the passage of carriages, 9 feet wide, and 15 feet high to the springing of the arches, the wall between them being a little more than 6 feet thick.

The masonry above the openings is tied together by iron dowels 1 inch diameter and 12 inches long, put through holes drilled through each stone of every course. And to prevent the walls of the arches for the carriage roads being pressed out by the weight of the superincumbent part of the pyramid, six wrought iron tie-bolts, 4 inches wide by 2 inches thick, are laid in the masonry at the springing of the arches, with dovetail heads at each end, which are let into corresponding sockets in cast iron plates bedded in the masonry.

The main piers are connected with the shores by a series of arches (see fig. 1.), viz., three on the Carnarvon side, and four on the Anglesey side. The height of each of the small piers, from high-water line to the springing of the arch, is 65 feet, and the span of each arch 52½ feet. Like the main piers, they are not built solid all the way up, but a square chamber is left in each, filled up with rubble.

73. *Main chains.* — The plan originally proposed had been the same as that for *Runcorn Bridge*, viz. to make iron cables of small bars welded together and bound up in a bundle. * This, however, was laid aside, and the chains were ultimately made on the plan brought forward by Captain Brown, viz. of straight bars, united by coupling bolts. †

† *Vide antè*, art. 68.
* The goodness of the system of combining small iron rods by welding them together, and thus forming them into cables,

Fig. 2. shows one of the bars, and one of the coupling plates.

The main chains are 16 in number, each containing a series of links, composed of 5 wrought iron bars, 10 feet long, $3\frac{1}{4}$ inches broad, and 1 inch thick. (See fig. 2.) They are disposed, four chains one under the other, on each side of the central foot-path, and four at each outside of the platform. There are, therefore, in all, 80 bars in the main chains, and their united section is $(80 \times 3\frac{1}{4} =)$ 260 square inches of iron. The bars are united by coupling plates 16 inches long, and 8 inches broad by 1 inch thick (see fig. 2.), and screw bolts 3 inches diameter, each bolt weighing 56 lbs.

It had been originally intended to pass the main cables over cast iron frames of a pyramidal form, erected on the main piers (see fig. 6. Plate III.), and thence down to another cast iron frame laid horizontally along the top of the masonry of the back arches,

---

may be deemed rather doubtful, notwithstanding the high authority of the engineers and scientific men who gave their opinion in its favour, before the Parliamentary Committee. (*Vide* Third Report on Holyhead Road, April, 1819.) Cables made of small wires have been used with success, but then the wires are of the whole length of the cable. In making up a cable of bars welded together, the difficulty of turning and handling a bar of iron of 500 feet, and upwards, in length, so as to make the joints sound, would be very great.

The adoption of the plan of making the chains of a series of straight bars bolted together, by Mr. Telford and by all the engineers who have built suspension bridges, is at any rate a sufficient authority for the student to prefer it, as the best known method for large constructions. It has, besides, a very certain advantage over the cable plan, by rendering the entire chain less liable to be deranged by the failure of any one particular part, and by giving greater facility for repairing any injury.

and connected with their springers by means of perpendicular rods, thereby embracing the whole mass
of masonry and spandrels; making in all a mass of
about 12,000 tons (exclusive of the great pyramids)
at each end of the bridge, to keep the ends of the
main chains down. *

This plan was laid aside. The main chains are carried through tunnels, cut through the solid rock of
which the shores are composed, and the extreme ends
are fastened by strong holding bolts, in chambers
made at the ends of those tunnels. The chains are
thus attached to the solid mass of rock of the shores,
and must pull that away before they can get loose
from their fastenings. The holding bolts are 9 feet
long, and 6 inches diameter, and rest in sockets in
cast iron plates 6 inches thick.

There are in all 12 holding bolts at each end of
the bridge, viz. 4 for the two middle sets of chains,
and 4 for each of the outer ones. The length of
each bolt, between its bearings against the rock, is 18
inches.

There are three tunnels for the chains, about five
feet square, at each end of the bridge, viz. one for
the two central sets of chains, and one for each of the
outer ones.

The bars of the chains in the tunnels are 7 feet
6 inches long, with their connecting plates; 4 inches
broad by 1½ thick, and the coupling bolts are 4 inches
diameter. They were made thus much stronger than
the other bars, under the idea that in the tunnels they
would be more exposed to oxidation, and less acces-
sible for painting or varnishing them.

The back stays are tied down, by vertical rods,

* See Mr. Telford's Report, in the Third Report of the Committee on the Holyhead Roads, 1819, p. 335.

(see fig. 1.) to small cast iron plates laid in the masonry of the arches, between the main piers and the shores, to prevent their undulation.

Thus the main chains are fixed quite fast, at their ends, into the rock, and held down also between the main pyramids and the shores. But over the pyramids they lie loosely upon cast iron saddles, which are themselves laid upon horizontal rollers or trucks, 3 feet 8 inches long, and 8 inches diameter, that lie in grooves or channels formed in a cast iron platform, fastened down upon the summit of each main pier. The saddles can thus move backwards and forwards a few inches in the direction of the length of the chain, and hence expansion and contraction of the chains produce no other effect than to move the saddles on which they lie, just as much as they lengthen or shorten, and to raise or depress the suspended roadway a little, instead of producing injurious strain on the materials. The rollers on which the saddles rest are kept equidistant, by their necks or projecting ends being received in brasses, held between wrought iron plates that are screwed together, so that no one roller can move alone; but there is sufficient room in the grooves in which the rollers are placed, for

The chains are tied together across the breadth of the bridge by *transverse ties*, which are cast iron tubes placed between each pair of chains, with wrought iron screw bolts $1\frac{1}{4}$ inches diameter put through them, and passing through the chains. The bolts serve to draw the chains together by their nuts, and the cast iron tubes keep the chains a proper distance apart.

There are 8 of these transverse ties in the length of the bridge, viz. 2 between the first and third lines of chains, and 2 between the second and fourth.

The chains are further tied together by a diagonal wrought iron lacing, between each pair of transverse ties.*

74. In the length of the chains there are 4 sets of *adjusting links* to every line of chains, viz. 2 in the opening between the piers, and 1 between each pier and the shore. The object was to give the means of adjusting the chains to their proper length, in case there should be any inequality in the lengths of the bars. The following figure is a perspective view

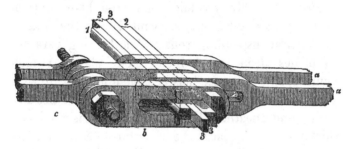

of one of the adjusting links, and fig. 3. Plate IV. is a side view of it on a large scale. At one end of each of the bars *a*, a long hole is made instead of a round eye, and the corresponding end of the connecting plate *b* has also a long hole in it. The bar and connecting plate are connected by two semicircular pins, 1 and 2, one bearing against the ends of the bars *a*, and the other against the ends of the connecting plates *b*, and the space between them is filled up with iron wedges, 3 3. If any one of these iron wedges is taken out, the ends of the bars *a* can be drawn away a little from the ends of the connecting plates *b*,

* This mode of binding the chains together to stiffen them, was adopted in consequence of the effect perceived on the chains from the action of gusts of wind, while the bridge was still unfinished.

and so the chain will be lengthened ; or by driving in
more wedges, the two semicircular pins, 1, 2, are se-
parated more, and the chains shortened. The bars *c*
are connected to the other ends of the connecting
plates *b* by screw bolts, like the other bars of the
chain. The whole distance that each adjusting link
can be lengthened is 9½ inches ; therefore the chains
can be lengthened 19 inches between the pyramids,
and 9½ more between each pyramid and the shore ;
that is, 38 inches in all.

75. The roadway is laid upon wrought-iron cross-
bars or *roadway bearers*, suspended from the ends of
the vertical suspending rods. The bearing bars are
3½ inches deep and ½ inch thick, placed 1 inch
asunder, and the space between them is filled up with
a piece of deal. They are trussed, with bent ties and
struts underneath. The *roadway* consists of two
thicknesses of fir plank, the lower one of 3 inches, the
upper one of 2 inches, laid parallel with the length of
the bridge, upon the iron cross bearers. The lower
course is bolted to the wood that is riveted between
the iron roadway bars, and is covered with Borrodaile's patent felt, saturated with boiled tar; the
second course of planking is spiked through to the
lower course.

In the middle of each carriage-way a third course
of planking, 2 inches thick, is spiked through to the
second, through an intermediate layer of felting;
and an oak guard is bolted on each side of the carriage-way through the two lower courses of planking.

Also, to stiffen the roadway, an oak plank is bolted
to the underside of it, between each of the cross
bearers.

All the suspended ironwork of the bridge was
proved by actual strain in a proving machine before

being put up, at the rate of 11 tons strain per square inch.*

The plan of operations in putting up the chain-work was this:—

76. The chains were first put together in the tunnels, working up from the fastenings to the mouths of the tunnels, by bringing down one link of each chain at a time, and bolting it to the one brought down before. Next the chains were put together from the mouths of the tunnels up to the main piers, upon a scaffolding erected on the masonry between the pier and the shore, with the proper inclination for the back stays. The chains were then continued over the Carnarvon pier, hanging down loose nearly to the level of high water. This was done by suspending a cradle large enough for two men to sit in, from a crane arm set up on the top of the pyramid. The cradle was suspended by tackle, so that the men could slack it down, or haul it up, to raise or lower themselves at pleasure. The links of the chain that were to be joined on to the ends of the chain hanging over the top of the pier, were brought successively along the road to the front, or seaface, of the pyramid, through the arched roadway opening. Thence the link was taken up by tackle from a pair of shears placed on the top of the pyramid, and lifted up to the height of the link it was to be fixed to, where the men in the cradle got hold of it, and brought the two links together, and put the coupling bolt through them.

On the Anglesey side of the strait, the chains were carried just over their saddles on the top of the pier,

* The construction of the proving machine was very nearly similar to that described in the note to art. 33.

and their ends were retained by tackle made fast to them, and descending thence to capstans on the shore. The remaining piece or length of chain, which was to unite the two ends of the chains brought up from each shore, was laid on a raft, 400 feet long, and 6 feet wide, made of whole deal baulks bolted together. One end of that piece of chain was first fastened to the end that hung down to the water from the Carnarvon pier, and then the raft was floated across to the Anglesey pier, and the loose end of the chain upon it was fastened to the tackle that hung down from that pier. The tackle was then hauled up by the capstans fixed on the shore, and the chain raised gradually off the raft, until the end of it was brought in contact with the end of chain that hung over the top of the pier. The two loose ends were then bolted together, and the operation completed. The first chain of the Menai Bridge was thus raised and fixed on the 26th of May, 1825, in one hour and thirty-five minutes. The other 15 chains were successively got up in the same way at subsequent periods.*

Previously to suspending the main chains, experiments were made by Mr. Rhodes to find the *tension* required to draw them up to the required curvature. They were tried with a chain made of some of the vertical suspending rods, 1 inch square, with a chord line of 570 feet, and deflections varying from 35 feet to 49 feet. He found the tension of the chain at 43 feet deflection, viz. the intended deflection of the chains, = 6134 lbs. The weight of the chain was

---

* For the dimensions and details of the machinery used in these operations, the author refers the reader to Mr. Provis's Account of the Menai Bridge.

3599 lbs.; hence the *tension* was 1·7 times the weight suspended.

77. The hoisting tackle consisted of two ropes, each leading from a capstan fixed on the shore between 400 and 500 feet back from the pyramid. Each rope led up to the summit of the pyramid, and there passed over two pulleys 14 inches diameter, fixed in a strong frame placed on the top of the pyramid. Thence the rope passed through a block of four sheaves, made fast to the end of the piece of chain on the raft. Thence the rope led up again through a three-sheaved block at the top of the pyramid, lashed to one of the chains of the back stays, and, lastly, through a single-sheaved block made fast to the end link of the chain hanging over the summit of the pyramid, to which the end brought up from the raft was to be attached. The barrels of the capstans were 1 foot 8 inches diameter; the axles about 4 inches diameter, worked by 8 spanners of 10 feet radius, and manned by about 150 men for the two.

78. The bridge was opened on the 30th of January, 1826, viz. 6½ years after its commencement; and the several improvements and alterations that it was found advisable to make in order to stiffen the chains and roadway were completed during the summer of 1826, since which the bridge has stood hitherto unimpaired, and in perfect security.*

There is now no perceptible transverse vibration of

* A detailed account of all the operations in the erection of the Menai Bridge, especially in raising the main chains, with accurate drawings of every machine and apparatus used in them, will be found in Mr. W. A. Provis's Account of the Menai Bridge. Folio. London, 1828.

the roadway. The vertical undulation before the introduction of the transverse ties, and diagonal lacing, had been (according to Mr. Provis) about 18 inches; but since, it has not been observed to exceed 6 inches. The chains are scraped and painted once in three years; and no part appears to have become deranged. The extent of longitudinal motion of the saddles on the pyramids by expansion and contraction of the chains has been observed to the extent of 2 inches, which shows the expediency of Mr. Telford's contrivance for allowing play for them.

79. The Menai Bridge is the largest, and hitherto the finest suspension bridge that has been erected. It has indeed been not unfrequently made the subject of praise, expressed in language bordering on exaggeration. Considering, however, the difficulties of the situation, and the few data that existed for such constructions at the time that Mr. Telford undertook it, it must be pronounced a specimen of great boldness of design, united with the utmost care and judgment in execution, and the student will improve both his knowledge and his taste by surveying it very attentively.

80. The system of proceeding adopted by Mr. Telford was that which is advisable for all great, and especially new, constructions. In the beginning nothing was fixed but the general outline of the plan, and this was not very essentially departed from afterwards. But all the details of execution, particularly of those parts on which there was very little previous experience, were left to be devised until the time arrived for their execution. Thus Mr. Telford was not shackled by the necessity of adhering to any complete settled plan, a slight departure from which might produce some derangement that it would have

been impossible to foresee. He gained also the advantage of longer thought, and of all the experience that he acquired as he advanced. And above all, he could keep his mind free ; and select from the ideas that occurred to himself or to his assistant engineers, without being biassed by any affection to preconcerted schemes. The result of this good system, combined with his great practical skill, has been the production of a work of art, undoubtedly, of the first order.

81. The weight of the 16 main chains between the points of support, including connecting plates, screw pins, wedges, &c., is computed by Mr. Rhodes, the transverse ties, the weight of the suspending rods and platform, &c.

	Tons.	cwt.	qr.	lbs.
81. The weight of the 16 main chains between the points of support, including connecting plates, screw pins, wedges, &c., is computed by Mr. Rhodes,	394	5	0	16
The transverse ties,	3	16	2	20
The weight of the suspending rods and platform, viz. roadway bars, planking, side rails, iron work, &c.	245	13	2	27
Making the total suspended weight	643	15	2	7

Whence the *tension* on the iron at each point of suspension, by the weight of the bridge itself, is, according to Mr. Rhodes's experiment (see end of art. 76.), 1·7 times 643 . 15 . 2 . 7 = 1094·42 tons.

The entire section of the 80 bars is (80 × 3·25 =) 260 square inches, which would bear without breaking (260 × 27 =) 7020 tons.

According to experiments made by Mr. Telford, about half the breaking strain, or 3510 tons, would produce permanent elongation. And if we take the standard of 9 tons per square inch, the chains will bear constantly without any risk (9×260 =) 2340 tons, or $2340 - 1094\cdot42 = 1245\cdot5$ tons of strain more than the strain produced by the weight of the bridge itself.

Therefore it may be perpetually loaded, without any injury, with $\frac{1245\cdot5}{1\cdot7}$, or about $732\frac{1}{2}$ tons besides its own weight.

CONWAY BRIDGE, ERECTED BY MR. TELFORD.

82. The Conway Bridge was built to replace a ferry across the estuary between the town of Conway and the opposite shore that leads to Chester, and to connect by a safe and permanent road across that estuary the towns of Bangor and Chester. It was begun in 1822, and completed in 1826 ; and its construction differs very little from that of the Menai Bridge. The principal change was in the manner of putting up the main chains, which was done on the plan first proposed for the Menai ; viz. a temporary rope bridge was made with the ropes used for the hoisting tackle at the Menai Bridge ; and on that the chains were put together in the places intended for them ; after which the ropes were slacked and the chains were adjusted to their proper tension.

The following are the principal dimensions of Conway Bridge : —

	Feet.
Distance between the points of support, -	327
Deflection, - - - - - -	$22\frac{1}{3}$
Height of under side of roadway above high water of spring tides, - -	15

Strength of the Chains. — There are 8 chains, each consisting of 5 bars $3\frac{1}{4}$ inches broad by 1 inch thick. The total section of iron, therefore, in the chains is $(40 \times 3 \cdot 25 =)$ 130 square inches, which will bear constantly, without injury, 1170 tons. The deflection $\frac{1}{15}$th of the chord line ; the tension for that deflection is equal to *twice* the weight. Whence the weight of

the bridge and its load must not exceed the half of
1170 tons, or whatever the weight of the bridge itself is, it will bear in addition the difference between
its own weight and 585 tons.

DESCRIPTION OF THE BRIGHTON CHAIN PIER, DESIGNED
AND CONSTRUCTED BY CAPTAIN S. BROWN, R. N.

(Figures 1, 2, and 3, Plate V.)

83. The Brighton Pier was commenced in Oc-
tober, 1822, and opened in November, 1823.

It runs out into the sea 1014 feet from the face of
the esplanade wall, and the entire length of the bridge
is 1136 feet, in four openings, each of 255 feet span,
and 18 feet deflection. The extreme breadth of the
platform is 13 feet, and the clear breadth 12 feet
8 inches.

The suspension towers consist of pyramidal cast
iron frames, one on each side of the bridge, united by
an arch at the top. They are 25 feet high, 10 feet
apart, and weigh each about 15 tons ; made of plates
riveted and cemented together with iron cement.
They stand on clumps of piles driven about 10 feet
into the chalk rock that forms the bed of the sea, and
rising 13 feet above high water. The clumps are
256 feet apart, centre and centre, and the clear openings between them 227 feet. The clump of piles at
the outer end, or pier head, is in the form of a T,
and contains 150 perpendicular piles, besides diagonal
piles, framed strongly together with walings and
cross-braces. On them a platform is laid, 80 feet by
40 feet, which is paved with a course of Purbeck
granite about twelve inches thick, and weighing
about 200 tons.

The three other clumps contain each 20 piles, be-

sides the diagonal piles, and are also braced firmly
together.

84. *Main chains.* — The platform is supported by
four chains on each side of the bridge, arranged two
in breadth, and two in depth. The chains are formed
of wrought iron round eye bolts, 6¼ inches in circumference, or about two inches diameter, 10 feet long
extreme length, and weighing 112 lbs. each ; united
by open coupling links, of iron 1½ inch deep by
1 inch thick, and bolt pins 2 inches diameter. The
total section of the iron of the chains is therefore
$3 \cdot 1416 \times 8 = 25 \cdot 1328$ square inches.

The chains rest upon saddles on the upper part of
the suspension towers. The chains on the land-side
are carried over a suspension pier of masonry, and the
back stays are carried through two tunnels in the cliff
between 30 and 40 feet deep ; viz. one tunnel for each
set of chains. At the end of each of these tunnels a
chamber of brick-work is built in cement ; and at the
bottom of the chamber a large square stone is fixed, to
which the ends of the set of chains are attached, and
they pass from it through another large stone placed
against the entrance to the chamber, at the back of
which they are finally fastened to a cast iron plate,
weighing about 1¼ tons. The chains thus take a firm
hold of the cliff itself.

The back stays at the pier head are fastened by
strong holding bolts to the diagonal piles of the clump.

85. *The platform* is suspended from the main
chains by vertical suspending rods 1 inch in diameter,
and 5 feet apart, viz. one at every coupling of the
chains. At their upper ends they are formed with a
cross or T head, (see Figures 2 and 3,) which is supported by a cap resting upon the ends of the main
chains and coupling links, one cap at every coupling.

A square cavity or chamber is cast in the lower part of each of the caps, with an oblong slit or entrance, through which the T head of the vertical rod is put upwards, and being turned round in it across the slit, is borne by the cap, as shown in Figure 3. The lower ends of the vertical rods spread out into forks, which clasp over, and are fastened by cross pins to a longitudinal side bearer made of bars of flat iron 4 inches deep by ¾ thick, bolted together and extending from tower to tower. On the side bearers are laid cross joists 7½ inches deep by 3½ thick; and over these is laid a course of 2½ inch planking for the roadway. The platform is guarded and stiffened by an iron railing or parapet.

86. Considerable difficulties were experienced in the execution of this work.*

In attempting to establish the first clump of piles, the scaffolding that had been erected upon the first piles was twice swept away by storms, and some of the piles torn up. The bed of the sea is chiefly chalk rock, but at intervals there are hard rocks : to prevent the piles from being driven upon these instead of into the chalk, it was found necessary to make the iron shoes of the piles much larger than usual, and they were bound to the wood by iron straps 3 feet long attached to the solid point, and bolted to the wood.

It was found also requisite, on account of the hardness of the foundation, to use more powerful pile engines than usual, viz. with monkeys weighing about one ton each.

When the first clump of piles was driven, a temporary rope bridge on the South American plan was

* The following particulars are taken from an account published by S. Simes, clerk of the works. 1830.

made, for the workmen to pass to and fro, by fixing on the deck laid down on the piles a couple of upright iron stanchions about 4 feet apart. Over these stanchions cables were carried, one end being fastened to the deck of the piles, and the other end carried away and fastened in the cliff. A square car was then suspended capable of holding four men, in which the workmen pulled themselves across.

To drive the second clump which was in deeper water, two pairs of shears were floated out at high tide, and then their legs sunk at low tide by fastening stones to them. On these two pairs of shears thus erected in the water, a scaffolding was made for the pile enginés. The piles of the second and third clump were driven in the same manner as the first set, and the difference in the operations being only that of the increased labour and perseverance requisite, there is nothing in them particularly worthy of being recorded. The temporary bridge of cables for the men was continued as the clumps were successively driven.

In beginning the fourth clump for the pier head it was found impracticable, owing to the depth of the water (viz. 14 feet at low tide), to fix the common wood shears. Other shears were therefore made of 4 inch cast iron pipes, put together with lead joints, four in a length, with a strong pole driven through them, each set of four pipes making one leg of the shears. Six of these legs were floated out to the place for the pier head one at a time, by lashing it to four casks; and one of the ends was sunk till it rested on the bottom, by cutting off three of the casks successively. The six legs were all thus sunk, and their upper ends being connected like common shears, a scaffold was built upon them. This plan did not answer, one of the pairs of shears being carried away by a storm, some of the poles of the other snapped, and the whole disabled so as to be unfit for service.

87. Other shears were then made, each leg of which consisted of bars of 1¼ inch iron, welded together in three lengths of 12 feet each ; making therefore a bar 36 feet long. Nine of these bars were bound together by iron bands, 2 inches wide, and 18 inches apart, to form one leg of the shears. The section of each leg, therefore, was 14 square inches ; and the weight about 3·32 lbs. (the weight of a bar of iron 1 foot long and 1 inch square) $\times$ 14 $\times$ 36 $= 1678$ lbs., or nearly ¾ ton.

These shears were floated out and fixed, and stood well.

The temporary bridge for the men was improved during the erection of the pier head, by laying a road of planks on the cables stretched from clump to clump, and making a sort of side railing to it, by other slighter ropes stretched a couple of feet above the cables, and fastened to them by wood posts lashed to both lines of ropes.

88. The strain that should be expected on a bridge placed in such a situation as the Brighton Pier, arises not so much from the actual load in dead weight that may come on it, as from the sudden shocks to which it is exposed. The former is rarely likely to approach the *maximum* that it will bear, and, except in the embarkation of troops, or some extraordinary ceremony or sight to be witnessed, a chain pier is never likely to be one quarter loaded.

But the strain that may be brought on the Brighton Pier, for example, by the effect of the violent tempests that prevail on that coast, is not susceptible of calculation ; and it appears almost singular that a structure apparently so slight should have escaped, unhurt, the storms that have assailed it. In 1824, for instance, a year after the erection of the pier, it stood through a storm which was violent enough to tear up trees.

Suspension bridges, notwithstanding their apparent slightness, are, in fact, peculiarly adapted for sea-coast service ; and will stand the shocks of wind and water better than more massive constructions ; because they present no large surfaces to those shocks, but are from beginning to end, as it were, a sort of net-work, the meshes of which allow the masses of water to pass through them without resistance ; so that the actual shock upon the materials themselves of such a bridge, when a heavy sea strikes it, is not great, although it appears to be struck with a violence that must overwhelm the whole structure.

In building a chain bridge for sea-coast service, the parts should be tied firmly together to make them resist vibration, that it may not be set swinging by gales of wind. But the whole structure should be made as open as possible, so that there may be *no large masses of surface* to receive the shocks of the wind and sea.

DESCRIPTION OF TWO BRIDGES OF SUSPENSION, CON-
STRUCTED BY M. J. BRUNEL, ESQ. F.R.S., AND ERECTED
AT THE ISLE DE BOURBON.*

(Plate V.)

89. These bridges were designed by Mr. Brunel,
and executed in England, near Sheffield, where they
were put together in January, 1823, before being
sent out to the Isle de Bourbon.

Fig. 4. is an elevation of the larger of the two; fig. 6.
a cross section through the platform; figs. 7. and 8. detached views of the cross bearers, on which the road is
laid; figs. 9. and 10. a side and end view of the central
suspension frame; and figs. 11. and 12. a side and
end view of one of the small suspension frames placed
on either shore.

The bridge, fig. 4., consists of two portions of
catenaries, suspended over a central frame about 25
feet high, placed on a pier of masonry in the middle
of the river, and over two end frames about $5\frac{1}{2}$ feet
high, erected upon land piers, or abutments, on
either shore.

There are, therefore, two openings, each of $131\frac{3}{4}$
feet span between the suspension bolts to which the
chains are attached; and the clear opening between
each land abutment and the central pier is 122 feet.

Under each opening is a set of 4 chains curved
upwards (see fig. 4.), and also sideways. These

---

* The following account of these bridges is translated from
Mons. Navier, who examined them in May, 1823, at Sheffield.

chains are fastened at their ends into the abutments
and the central pier, and are connected to the roadway by ties that are drawn up tight; the object being
thereby to stiffen the platform, and cause it to resist
the vibration produced by the violent hurricanes to
which these bridges are exposed at the Isle de Bourbon.

90. There are three sets of main chains, placed at
9 feet 8 inches apart, centre and centre, leaving thus
two roadways, about 8¾ feet wide each. Each set
contains two series of open wrought iron links, 4 feet
8 inches long, inside measure, made of round iron
1·36 inch diameter. Therefore there are, in all, six
sets of links; and the total section of iron in the
chains is 17·4 square inches. The long links are
connected by short coupling links 8¾ inches long inside, made of iron 1·36 inches by 1 inch. The bolt
pins are 2 inches diameter.

One of the coupling bolts goes through both the
chains of each set, and supports one of the vertical
suspending rods, which are made with eyes at their
upper ends for that purpose, through which the coupling bolts pass. The bolts are not held by round heads
and keys, as usual, but a half head is made at each
end of the bolt which overlaps the end of the link, so
that the bolts cannot draw out lengthwise, but yet
can be put into their places through the open parts of
the links.

An adjusting link is put in next to each suspension
frame, and at every fourth link therefrom, by making
one of the coupling bolts in two pieces, with a wedge
driven in between them. By taking out that wedge
the chain is lengthened; or by putting in several
wedges, it is shortened, in the way described of the
Menai Bridge, antè, art. 74.

The vertical supending rods are 5 feet apart, measured on the chains, and are made of round iron, $1\frac{1}{4}$ inches diameter. The four first suspending rods from the central pier are made in two lengths, united by a link (see fig. 9.), through which a horizontal rod (see figs. 4. and 9.) passes. One end of that horizontal rod is attached to the central suspension frame by a screw link, to tighten it up or lengthen it; the other end is attached to one of the coupling bolts of the sixth link of the chain. (See fig. 4.)

91. The road is laid upon cast iron cross bars, supported by the vertical suspending rods. These cross bars are T-shaped (see fig. 8., which is a section of one of them); viz. they are flat plates 4 inches broad by $\frac{3}{4}$ thick, with a vertical rib cast on their under sides, $5\frac{1}{2}$ inches deep at each end, and 7 inches deep in the middle, the lower edge of that rib swelling out to a rounding bead 2 inches broad.

The cross bars are made in two lengths, and when in their places under the platform, the outer end of each is borne by the suspending rod of the *outer chain*, while the suspending rod belonging to the *middle chain* goes through the joint of the two cross bars. (See fig. 6., which is a cross section of the middle part of the platform, to show one of the iron cross bearers.) The outer end of each cross bar swells out into a cylindrical socket, through which the lower end of the outer suspending rod passes, and is screwed up under it with a nut.

The opposite or inner ends of the bars, where they join together under the middle chain (see fig. 6.), swell out each to a half cylindrical socket, with a flange cast on it (see fig. 7.), and the ends of the bars are screwed up together with a couple of screw bolts passing through their flanges, so as to form, when the bars are united, a cylindrical socket for the central supending rod. (See figs. 6. and 7.)

The flat tops of the cross bars spread out at each end to a plate 8 inches square (see fig. 7.) ; and on these square plates a line of timbers, 8 inches broad by $8\frac{1}{4}$ deep in the middle, and $5\frac{1}{2}$ at the sides, is laid lengthwise of the bridge, from bar to bar, under each main chain. In fig. 6. the central longitudinal timber is shown in section. These longitudinal beams are fastened down to the cast iron cross bearers by two bolts passing through an iron cap laid over the upper side of the beam ; and a hole is made through that cap, and through the beam, for the vertical suspending rod to pass through. (See figs. 6. and 7.).

The roadway consists of three courses of longitudinal planking, laid on the cast iron cross bearers, viz. one course for the carriage wheels, 12 inches broad by 4 deep ; another for the horse track, 2 deep ; and a third, $1\frac{1}{4}$ deep, for the footpath, laid upon the course for the horse track, close up against the large longitudinal beams. The wheel track is covered with longitudinal rails of wrought iron $\frac{3}{8}$ thick, and the horse path by transverse rails $\frac{1}{4}$ thick. The planking is bolted to the flat tops of the cross bearers. *Note,* all the wood work of the platform is of *teak wood.*

The parapet is formed by vertical standards $\frac{7}{8}$ diameter, placed in front of each suspending rod. The lower ends of the standards are keyed into cylindrical sockets, cast out of the caps that cover the longitudinal beams, and the upper ends are held in clamps or bridles, bolted round the vertical suspending rods. (See fig. 9.) A railing is attached to these vertical standards, formed of an upper and lower longitudinal rail 2 inches broad, with small vertical rails about $\frac{5}{8}$ diameter.

92. The platform in each opening of the bridge is
inclined downwards from the main central pier to the
abutment, the top of the former being about 4¼ feet
higher than the top of the latter. The main chains
form a curve whose highest point (viz. the point
of suspension over the main central pier) is 25½ feet
above the top of that pier; the points of suspension
over the land piers or abutments are 5¼ feet above
the tops of those abutments; and the lowest part of
the curve is 11¾ inches lower than these latter points
of suspension.

The under tie-chains are made each of a set of
iron rods 1¼ inch diameter, with eyes at their ends:
they are connected by coupling plates 11¾ inches long,
3¼ inches broad by ½ an inch thick, with bolt pins
1¼ inches diameter put through them. (See fig. 11.,
where a part of one of the tie chains is shown.)

A hole is made in the middle of each pair of
coupling plates, to fasten to them, by a bolt pin 1¼
inch diameter, the lower end of one of the rods that
tie the platform to the under chains. Those rods go
up in an inclined direction through the main longitudinal bearers of the platform (see fig. 6., where
two of them are shown), and they are held to them
by nuts.

93. The framing of the main standard of suspension in the middle of the bridge is shown in figs. 9.
and 10., and consists of the following parts:—

1st. A bed plate (a a fig. 10.) is laid on the masonry
of the pier.

On it are laid, crosswise, three square cast iron
troughs, or inverted boxes, shown in figs. 9. and 10.

These square boxes receive the ends of the main
longitudinal beams of the platform; and upon them

stand three triangular standard frames *b b*. (See figs.
9. and 10.) The whole being bolted down upon
the pier by long foundation bolts, which go through
the feet of the triangular standards, through the
flanges of the boxes, and through the bed-plate, and
are keyed under cast iron holding plates laid in the
masonry of the pier.

A broad flat plate is cast at the top of each leg of
the triangular standards *b b*. Those plates serve as a
bed for the square ends of three cast iron pipes, *c*, to
lie upon (see figs. 9. and 10.) ; and a second set of
triangular standards is erected on the square ends of
the three pipes *c*.

Top plates are also cast on the upper ends of these
triangular standards, to which a cross frame, *d*, is
bolted, which has an open part, or socket, cast in it
over each standard, to receive one set of chains. (See
fig. 10.) Each of the main chains is attached by a
bolt pin to the lower end of a vertical link, suspended
from a bolt pin which is put through the upper part
of the socket of the fixed cross frame. (See figs. 9. and
10.) Hence the chains can play a little lengthwise,
to allow of expansion and contraction. The framework of the suspension standard is all bolted together
by short bolts, which pass through the respective
flanges of the several parts, and is further tied together by long bolts; viz. two that tie the upper
middle cast iron pipe, *c*, to the cross frame, *d*, and
four that tie the two outer cast iron pipes, *c*, to the
same upper frame, and four bolts that tie the
two outer cast iron pipes, *c*, to ears, *e*, on the bed
plate, *a*. It is needless to give a more minute description of all these parts, as they are shown by the
drawings in Plate 5.

The frame-work of the iron standards placed on

the abutments is shown in figs. 11. and 12. Fig. 11. a side view of one of the frames, and fig. 12. an end view.

The chains are suspended from them by vertical links, in the same way that they are suspended from the central frame. The back-stays (*i*, fig. 11.) are formed of bars 3 inches broad by 1¼ thick, and 10 feet long, united by coupling-links, and by bolt-pins 2⅜ inches diameter. The one next to the suspension frame is an adjusting link.

The extreme ends of the chains are fastened to large circular holding-plates, 3 feet diameter and 2⅜ inches thick, strengthened underneath by four ribs 5¾ inches deep. These plates are sunk deep in the ground, and loaded sufficiently to keep them firm in their place.

94. The *second bridge* (see fig. 5.) is of one arch 131¾ feet span between the suspension bolts, which are 15 feet 5 inches above the tops of the piers on which they stand, and the deflection of the main chains is about 9 feet 7 inches. The breadth of the platform is the same as in the bridge (fig. 4.) ; and the general arrangement and construction are also similar to that of the first bridge, and therefore require no explanation.

DESCRIPTION OF THE HAMMERSMITH SUSPENSION BRIDGE,
DESIGNED AND ERECTED BY W. TIERNEY CLARK, ESQ.,
CIVIL ENGINEER.

(Plate 6.   Figures 1. and 2.)

95. This bridge is built across the Thames, eight
miles above London, communicating between Hammersmith and the Surrey shore of the river.   It was
commenced in the summer of 1824 and opened
in 1827.

The chord line or distance between the points of
suspension is 422 feet 3 inches.   The deflection of
the chains is 29½ feet, or $\frac{1}{14\cdot3}$ of the chord line ; and
the tension on the iron at the points of suspension is
1·857 the times the entire weight suspended.*   The
total water way is 688¾ feet, viz. 400¼ feet through
the middle opening, 145½ feet between the Surrey
Pier and the bank, and 143 feet between the Mid-
dlesex pier and the bank.

96. *The main piers* are of stone, designed as
arches of the Tuscan order.   They are 48 feet high
above the level of the roadway, 22 feet thick by 46
wide at that level, and 72 feet wide at low water line.

The basements of the piers are built of large blocks
of stone, ranging between 3¼ feet and 4 feet in
length by 1½ feet to 2 feet in thickness.   Through
each pier an arched opening is formed for the roadway 14 feet wide.

The platform is divided into a carriage way in the

* By Formula V. for computing the tension, *post*, Section VIII.

middle 20 feet wide, and two footpaths, each 5 feet wide.

97. *The main chains* are eight in number, arranged in four double lines, viz. two small chains, one under the other on the outside of each footpath, and two large chains on each side of the carriage way.

Each of the small chains consists of three lines of bars, 8 feet 10 inches long from centre to centre of the eyes or bolt holes, 5 inches broad and 1 inch thick. The ends of the bars are united by screw bolts, $2\frac{5}{8}$ inches diameter, put through coupling-plates, $15\frac{1}{4}$ inches long from centre to centre of the bolt-holes, 8 inches broad and 1 inch thick.

The large chains contain each six lines of bars of the same dimensions as those of the small chains, linked together also by bolts $2\frac{5}{8}$ diameter and coupling-plates. The chains contain, therefore, in all 36 bars, each containing 5 square inches, whence the total section of iron in the main chains is (36×5) 180 square inches. The chains pass through openings in the main piers, in which are placed roller carriages for the chains to rest upon. These carriages are cast-iron frames of metal 1 inch thick, with ribs cast on them to stiffen them (see fig. 2. plate 6.); and they stand on a bed plate let in and bolted down to the masonry of the pier.

Each carriage carries two sets of rollers one under the other, one for the upper and one for the lower chain of each double line. The rollers are cast iron ; their necks are of wrought iron, and rest in brasses in plummer blocks cast on the carriages. (See fig. 2.) The links that pass over the points of suspension are stronger than the other links, and are curved to lie on the rollers, which are arranged in a circle of such curvature, that the direction of the main chains is a tangent to it.

98. From the tops of the suspension piers, the back stays pass down at the same angle as the chains of the central opening to the abutments, which are built on the shores, of brickwork faced with stone.

The abutments are 45 feet long from front to back, 40 feet wide, and 13 to 15 feet deep from the top to the foundation. Making the absolute weight of materials of each abutment 2160 tons to resist the draw of the chains, exclusive of its resistance of adhesion.

The chains are carried through tunnels in the abutments, 2 feet wide for the small chains, and 3 feet wide for the large ones. The end links pass between strong cast iron holding-plates that bear against a large surface at the back of the masonry, and they are fastened at the backs of those holding-plates by very strong cross pins.

99. The chains are of hammered iron, made of thin flat bars welded together under a forge hammer, until the entire length and shape was produced without a shut. The eyes, or bolt holes, at the ends, were drilled out of the solid; and the ends of the bars swell out to the same depth as the coupling-plates, viz. 8 inches. The bars were bored in parcels of 6 and 3 links put together, and proved at the rate of 45 tons for each bar, or 9 tons per square inch.

The vertical suspending rods are 1 inch square, and about 5 feet apart. They are attached to the coupling-plates of the main chains by short links, and screw bolts 1 inch diameter, which pass through the coupling-plates.

100. *The platform* consists of cross joists, 12 inches deep, 4 inches thick, of Memel timber, supported

by the vertical suspending rods; and on these are laid longitudinal beams, and courses of strong planking. (See figs. 1. and 2. of the following sketches.)

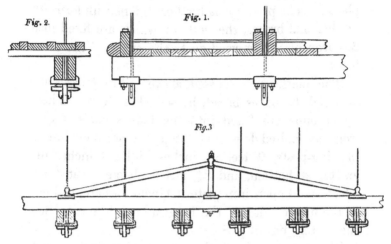

Each suspending rod passes down between a pair of the cross beams, which are 32 feet long, out and out. The suspending rods are thickened to $2\frac{1}{4}$ inches square at the lower ends below the cross beams, and pass through a flat cast iron plate, about 10 inches square, placed against the underside of the wood, with ribs or edges upon it that embrace the cross beams, and the ends of the suspending rods are fastened under these plates by keys and keepers. (See the sketch, fig. 2.)

A line of longitudinal beams, 8 inches deep by 4 thick, is bolted down upon the cross joists outside of each footpath; and two other lines of beams of the same dimensions, on each side of the carriage way. The space between them is filled up, first by a course of longitudinal 3 inch planking. (See figs. 1. and 2.) Over this is laid Borrodaile's felting, saturated with pitch and tar; and over that a course of alternate

3 inch planks, 1 foot wide ; and timber 4 inches wide (laid endwise of the grain upwards) and 3½ inches deep, projecting therefore ½ an inch above the planks. The platform is further stiffened underneath by diagonal braces, the ends of which are fixed into cast iron plates, screwed on to the sides of the cross beams.

The platform is trussed above. The trusses are of wood, 6 inches broad, by 4 inches deep. Their upper ends are fitted into the hollow sides of cast iron caps, bolted down on the tops of cast iron columns or king-posts, 2 feet 2½ inches high ; 4 inches diameter at bottom, and 3⅜ inches diameter at top ; which stand upon iron plates 10 inches square, bedded in the upper sides of the longitudinal beams. (See the sketch, fig. 3. p. 85.)

The lower ends of the trusses abut against cast iron steps, bolted down on the longitudinal beams.

The king posts are held down by long bolts, which go up between the longitudinal beams, with a large washer plate under them. (See the sketch, fig. 3.) And the steps for the lower ends of the struts are bolted in a similar way to the cross joists, by the same bolts which secure the cross beams and longitudinal beams together.

The king posts are 25 feet apart, with 5 vertical suspending rods between each pair.

The sides of the platform are guarded by parapets, formed of vertical cast iron columns 20 feet apart, bolted down to the platform by the same bolts which secure the longitudinal beams outside of the footpath to the cross beams.

The standards are tied together by a horizontal wood rail, 3½ inches deep by 2 wide, laid along their tops, and by diagonals of wood of the same dimensions ;

and further by a horizontal lacing, of half round iron, bolted to the diagonals of wood and to the iron columns.

The ends of the cross joists are covered with an iron cap, to give them a neat appearance.

101. The back stays dip below the level of the platform between the piers and the abutments, as is shown in fig. 1. plate 6.; and therefore at that part it was necessary, in order to keep the platform horizontal, that it should be erected upon the chains, instead of being suspended from them.

This was done by bolting vertical cast iron frames on the chains at the couplings, and laying the cross joists of the platform on their upper ends, the height of the frames increasing, of course, with the dip of the chains below the platform.

The following sketch is one of the longest frames. It consists of a sort of chair (a) placed upon the

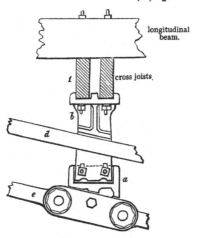

coupling-plates, and bolted through them. Another plate b, which has ribs cast on it to stiffen it, is bolted sidewise against the chair a. And the bolt holes are long slits, to allow the plate b to be raised or lowered a little in fixing it, in order to adjust its height to that of the platform above the chains.

At the upper end of the frame b, a square plate is cast, with three ribs on its upper surface. The ribs rise up outside and between the cross joists 1 of the platform, which are bolted down upon the plate, as shown in the sketch.

The vertical supporting frames stand on the lower chain e of each double line d e.

The cross joists and longitudinal beams being thus bolted down upon frames erected on the chains on each side of the platform, support it very steadily.

CHAIN BRIDGE AT BATH.

102. *Chain Bridge at Bath.* — This is a small bridge for foot passengers over the Avon, erected in 1830 by T. Shaw, Esq.

The platform is supported by two single chains made of wrought iron eye bolts, 5 feet long, and $1\frac{3}{4}$ inches diameter.

The chains are carried over four cast iron suspension columns at each end of the bridge; viz. two for each chain, lining with the length of the bridge, and terminated at the top by an entablature. The height of the points of suspension above the platform is about 14 feet. The span or horizontal distance between the points of suspension is 118 feet; and the deflection in the middle about $10\frac{1}{4}$ feet.

103. The suspension columns are about 16 inches diameter outside at the bottom, with a base 19 inches square; and they stand on a cast iron bed-plate, laid on a pier of masonry built of Bath stone, about 17 feet wide by $8\frac{1}{2}$ deep to the ground. The foundations of the piers are 10 feet deep, and built on piles. The links of the main chains are connected by coupling-links or plates $\frac{7}{8}$ thick, and 1 inch deep, united to the eyes of the long links, by bolt pins $1\frac{3}{4}$ inch in diameter, and 6 inches long. The coupling-links are formed with eyes in the middle, to which the vertical suspending rods are attached by screw bolts $\frac{5}{8}$ inch diameter. There are 23 vertical suspending rods $\frac{3}{4}$ inch diameter and 5 feet apart, between the suspension columns.

104. The platform consists of two longitudinal

roadway bearers of wood, supported by the vertical suspending rods, which pass through them, and are fastened underneath them with cross keys and keepers, an iron washer-plate, about ⅜ inch thick, being interposed between the wood beam and the iron key. The side bearers are made of two lines of beams each, about 4 inches broad by 10 inches deep, bolted together with screw bolts, about ⅝ inch diameter, at intervals of 2½ to 3 feet. They run the whole length of the bridge, but in several lengths, being scarfed at break joint, and the scarfing joints secured by iron plates and screw bolts. On the side bearers cross joists are laid, about 8 inches broad, 7 inches deep at the ends, and 9½ inches deep in the middle, and on these planking is bolted for the roadway. The platform is guarded and stiffened at each side by a parapet, consisting of pillars 2⅝ inches diameter, fixed on a longitudinal beam that is bolted to the platform over the side bearer. The pillars are 10 feet apart ; the upper rail of the parapet is of wood, 3 inches square, and is bolted to the tops of the pillars. The pillars and upper and lower rails are tied further by two diagonal wood braces 3½ inches by 3, between each pair of pillars, and by three horizontal rails 1¼ inch square, forming altogether a framework of considerable strength and stiffness. There is, nevertheless, a great vibration in this bridge.

The back stays are carried back at an angle of about 28 degrees, with the horizontal chord line, and are fastened in abutments of masonry built upon arches.

The iron work was executed by Captain Brown.

The bridge is warranted to carry 12 tons in load.

Calculation of the Bridge.

105. The two chains contain 4·8 square inches of iron, and will therefore bear a strain of (4·8 × 9 or) 43·2 tons before stretching.

The weight of the bridge, viz. of the chains, and other iron work, platform, and parapet, may be taken at about 7 or 8 tons ; the total weight therefore suspended with an additional load of 12 tons, would be say 20 tons.

The deflection of the chains is about $\frac{1}{13}$th of the chord line, and the tension at the points of support is about 1·4 times the weight. Therefore, 20 × 1·4 = 28 tons for the tension on each point of support.

If this bridge were entirely covered with people, its platform, which contains 118 feet × 9·5 feet = 1121 square feet, would carry at the rate of 2 persons per square foot weighing 140 lbs. per head, (560 × 140 =) 78,400 lbs. = 35 tons. Adding to that the weight of the bridge 8 tons, we have 43 tons, which × 1·4 = 60 tons for the tension on the chains at each point of suspension, if the platform were entirely filled by a crowd of people.

The bridge is, however, not likely to be ever loaded with more than the weight it is warranted for, as it is private property, and carriages are not allowed to cross it ; and moreover it does not communicate with any great thoroughfare.

DIMENSIONS OF A CHAIN BRIDGE AT BROUGHTON, NEAR
MANCHESTER.

106. The span is 145½ feet, and the deflection
12½ feet, $\frac{1}{11.6}$th of the chord line. The platform is
143 feet 3 inches long, and 18 feet 3 inches wide, and
is supported by four chains, two on each side of the
bridge, formed of round rods 2 inches diameter, and
4½ feet long, united by open coupling links of iron 1
inch square, and bolt pins 2 inches diameter.

The vertical suspending rods are round iron, 1 inch
diameter, and they spread out at the top into a fork,
connected by a bolt pin ⅞th inch diameter, to a plate,
which is interposed between each pair of the links
of the chains, and is secured to them by the coupling-
bolts.

The main chains are supported by four cast iron
suspension frames, viz. one at each end of the bridge
for each pair of chains. They are made of plates
bolted together with ⅝th inch screw bolts, at intervals of about 18 inches ; the frames stand on a base
of about 5 feet square, and thence rise up pyramidally.
The chains are attached at the points of suspension in
the way described (p. 80.) for Mr. Brunel's bridges ;
viz. they are connected by a pin to the lower ends of
a pair of vertical links, the upper ends of which are
supported by a fixed pin in the upper part of the
cast-iron suspension frame. Hence the points of
suspension are movable, to accommodate themselves
to the expansion or contraction of the chains.

107. The Broughton Bridge was broken down on
the 12th April, 1831, by a party of soldiers marching

over it in step. Sixty men are said to have been on
the platform at the time the accident occurred, their
line extending nearly the whole length of the bridge.
The failure was in one of the coupling-bolts that
held the ends of the back stays to their fastenings.
The last link of each chain was bolted to an iron strap
about 3½ inches broad, by a bolt 2 inches diameter;
and it was one of these bolts that was unable to resist
the violent effect of the vibration produced by the
march of the men.   The bolt is said to have been of
bad iron, the fracture presenting a granular and
crystalline appearance like cast iron.   The bridge
was quickly repaired, and is now in the state described above.

108. This accident shows the imminent danger to
a suspension bridge from bodies of troops marching
in step.   For although the bolt which broke, was not
as strong as in prudence it ought to have been, still
it was strong enough to have resisted a much greater
weight in *dead load* than it had to bear.

The weight of the bridge itself is about 43 tons*;
and the weight of the men on it, if *at rest*, was about
$(60 \times 180) = \frac{10800}{2240} = 4\cdot8$ tons; in all say 48 tons. The
tension for a deflection of 1 in 11·64 is about 1·55.
Therefore the utmost strain the holding bolts could
have had to endure, had the bridge been loaded with
60 men at rest, would have been $(48 \times 1\cdot55 =)$ 74·4
tons, or 37·2 tons for each bolt.

Now the length of the holding bolts between their
bearings was about 3·75 inches, and their diameter 2
inches, and each if of good iron would have borne

---

* The author is not aware of the exact weight of the bridge.
The above of 43 tons is computed from data communicated to
him by an engineer who had examined and measured it.

before being crippled, $\frac{\text{cub. } 2 \times 6000 \text{ lbs.}}{0\cdot312 \text{ feet}} = \frac{48000}{1\cdot312} = 153,846$ lbs. $= 68\frac{3}{4}$ tons.

Whence it must be inferred that the shocks, and the vibration given to the bridge by the simultaneous stepping of the men, must have produced an extra strain of about 30 tons beyond that of the mere load, in order to break the holding bolt.

A holding bolt of the dimensions above stated ought not, however, in prudence, to have been exposed at any time to a strain of more than 18 tons, had it been perfectly sound in the beginning; for under that strain it would be, if not actually deflected, at least on the very verge of deflection, and if so strained frequently, and for any length of time, the iron would probably become, by degrees, permanently injured. The holding bolts of the Broughton Bridge were, in fact, of such proportions that they would be perpetually exposed, in the common traffic of the bridge, to more than the strain they were capable of undergoing for a short time only, without receiving injury.

They ought to have been about $3\frac{1}{2}$ inches diameter to have borne as much as the chains would bear.

MICKLEWOOD BRIDGE, DESIGNED BY JAMES SMITH, ESQ.
OF DOUNE.

109. This bridge deserves notice from the platform
being supported upon iron frames standing upon the
chains, instead of being suspended from them.

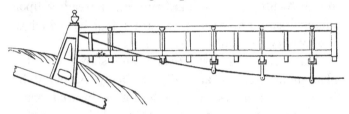

The span is 103 feet. The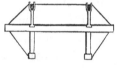
banks are very soft black mud,
and so bad for a foundation, that
it was judged inexpedient to
make abutments of masonry. The chains are, therefore, fastened to iron frames, (see the figure,) which
are bolted down upon a wooden framework, consisting of beams of wood laid in the mud at a considerable depth, not shown in the sketch.

On these, uprights are fixed, and over these again
cross beams, the whole being strengthened by cross
braces, and well bolted together, and earth and
rubbish are filled in and rammed down hard between
and around the beams of the frame. There are two
supporting chains on each side, 1 foot apart, and about
2⅜ inches diameter. The cross joists that bear the
platform rest upon iron uprights set up on the chains,
the lower ends spreading out like a fork to embrace
them. At every alternate joist the feet of the uprights

are tied together by an iron rod 1 inch diameter, extending horizontally under the bridge, and also by oblique tie bolts going up to the outer ends of the cross joists : and two other oblique tie bolts go up from the ends of the cross joists to the upper rail of the parapet. Further, where the chains lie above the platform, they pass through upright iron frames bolted to the cross joists, and to the top rail of the parapet. Hence, the parapet, iron standards, and cross joists are all firmly tied together, and stand steadily upon the chains. The cross joists are 1 foot by 5 inches, and twenty in number.

This bridge is very steady, and shows that the plan mentioned in art. 50., as being proposed by Mr. Stevenson, is practicable for a limited span. The stability is, however, obtained by excessive strength in the framing of the platform, and could hardly be obtained for a larger bridge without a prejudicial increase of weight.

SECTION IV.

FOREIGN BRIDGES.

PARIS SUSPENSION BRIDGE.

110. In 1823, Monsieur Navier, *Ingénieur en chef des ponts et chaussées*, proposed to the French government a design for a suspension bridge over the Seine, opposite the Hôtel des Invalides. The opening between the quay-walls is (150 mètres) 492 feet. The suspension pillars were to be built on the banks, and the distance between the points of suspension to be (170 mètres) 557½ feet. The distance from the points of suspension to the fastenings (32 mètres) 105 feet; deflection $\frac{1}{15}$th of the chord-line. The suspension pillars were to be four independent columns of stone, standing on piers of masonry built on piles. The columns were to be 10¾ feet diameter at the bottom, and 8¼ feet at the top. The piers were to be 13·12 feet broad at the top, and 19·68 feet broad at the bottom.

111. The suspension columns were to be tied transversely to each other by hollow square cast-iron pipes, 25½ feet long. The ends worked into the columns, to prevent their moving towards each other; and in the inside of the tubes were to be wrought-iron tie-bolts, the ends of which would be attached

H

to the columns, to prevent their moving away from each other.

112. The *main chains* were to consist of nine ranges of long open links on each side of the bridge, three in breadth, and three in depth.

The links to be of malleable iron, 16 feet long; and the iron 3·15 inches by 1·57 inches; united by coupling links of wrought-iron, 1¾ by 1·57 inches, and bolt-pins 4 inches diameter. The total section of iron in the main chains, 178 square inches. Distance apart of the chains 31 feet; breath of road between the parapets, 28½ feet; viz. a carriage-way, 18½ feet, and a footpath of 5 feet on each side. The backstays were to be carried back from the suspension columns through a stone pedestal, and then turned over an arched cast-iron plate, and carried down vertically into a well built of masonry, passing through blocks of stone bedded in the masonry-walls of the well, and fastened underneath a cast-iron plate, bearing against the undersides of those stones. (See the annexed sketch, fig. 1.)

113. The upper part of the masonry of the pedestal and well rested upon the stones (a); but as that superincumbent weight (says Mons. Navier) was not equal to the strain the main chains would have to bear, inverted arches of masonry (b, fig. 1.) were added, spreading out into the ground over a surface 60½ feet by 31 feet, surrounding the two wells. The weight of the ground upon these arches, and the adhesion of the ground to the masonry, were expected to add sufficiently to the resistance of the abutments. Also (says Mons. Navier), a stay, or buttress built on piles, was placed in front of each well, in a position to resist the pressure that is exerted (at the point where

the direction of the chains is changed) in the direction of the resulting force of the tension of the two parts of the chain. (See *d*, fig. 1., and Navier, p. 203.)

Fig. 1.

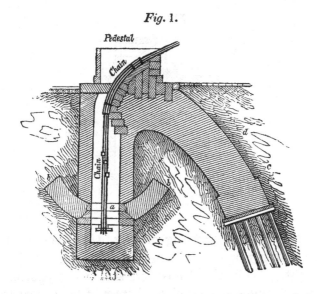

Pedestal

Chain

Chain

The entire weight of the suspended part of the bridge was computed 584,432 kilogrammes=129,159 lbs., or 577, say 580 tons.*

The estimated expense was 1,000,000 francs = 40,000*l*. The building of this bridge was begun in August, 1824. In the execution, an alteration was made in the abutments, by substituting for the spreading arches *b* (fig. 1.) a vaulted base, or found-

* The probable load on the bridge, in passengers, would be about 90 to 100 tons; and the possible load about 400 tons. Whence, as the tension for a deflection of 1 in 15 is about 1·94 times the weight suspended, the tension that the chains of this bridge would probably have had frequently to bear would be (580 + 100 =) 680 × 1·94 = 1319 tons. The greatest tension about 980 × 1·94 = 1901 tons; and the tension without any thing whatever on the bridge, would be 580 × 1·94 = 1125 tons.

ation (see fig. 2.), forming a base of 62 feet by 31 feet. The masonry being tied together by four iron tie-bolts lengthwise, and four others crosswise.

Fig. 2.

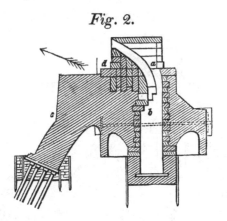

114. By the beginning of July, 1826, the chains had been got up on scaffolding, which supported their weight. While the chains were still thus supported, the tightening wedges of the first set of links, next to the entrance of the well, were driven up to begin tightening the chains. This operation produced an opening in one of the vertical joints of the courses of masonry at the upper part of the abutment, at (*a*), see fig. 2. The scaffolding was struck on the 30th of July, 1826, and then the weight of the chains increased the opening in the masonry of the abutments. When the joists that carry the platform were hung on the chains, the opening again increased; and, by the end of August, when nearly the whole of the platform was completed, and about $\frac{13}{15}$ths of the weight suspended, the vertical openings in the masonry of the four wells were about five centimètres = 2 inches (at *a*, fig. 2).

On the 6th of September, a water-pipe, which was

laid in the ground on the side of the Champs Elysées, and came near the abutments on that side of the river, burst, and, softening the ground about the abutment, it gave way under the strain produced by the weight of the bridge. The part of the masonry on the left of a b (see fig. 2), and above b c, was carried away from the remainder, and the suspension columns on the Champs Elysées' side of the river were pulled over towards the water.*

115. In consequence of this accident, and of the disputes that arose upon it as to the expenses of repairs, the undertaking was abandoned for the time, and the bridge removed.

Another bridge was afterwards erected nearly on the same spot, of which the following is a description.

* The defect of this abutment is obvious. 1st, It was not any thing like sufficiently massive and heavy in itself. 2dly, It was adapted to resist only the *vertical pull* of the chains; but their real pull was about in the direction of the arrows (fig. 2.). The buttress in front would therefore have little or no effect to resist the *effective pull* of the chains, and there was nothing to resist that pull but the weight and adhesion of the part b, c, d, of the masonry, aided by the resistance of the ground in front of it. The abutments of this bridge, according to the load it was likely to have to bear, viz. from 1300 to 1900 tons, ought not to have been less in *absolute weight*, at each end of the bridge, than from 4000 to 6000 tons, according to the depth of the foundations, and the nature of the ground; and the chains ought to have gone right through that mass, with large holding plates at the back.

(Fig. 3. Plate VI.)

116. The piers of suspension are archways built of stone ; they are 39 feet 8 inches broad, and 15 feet 5 inches thick at the level of the roadway, and rise up thence perpendicularly. Below the platform, the starlings spread out to about $54\frac{1}{2}$ feet broad, and 16 feet thick. The piers are built in the river, like the piers of the Hammersmith Bridge, which appears, in fact, to have been the model for the Pont des Invalides.

Through each pier is an arched opening for the carriage-way, 19 feet wide. The horizontal distance between the centres of the piers is $236\frac{1}{2}$ feet; and the deflection of the chains is about $\frac{1}{9}$th of the chord-line, or $26\frac{1}{3}$ feet.

117. The main chains are formed of round rods $1\frac{13}{16}$ inches diameter, 15 feet 8 inches long from centre to centre of the eyes; united by coupling plates 10 inches long from centre to centre of the bolt-holes; 4 inches broad, by $\frac{3}{4}$ inch thick. The bolt-pins $2\frac{1}{4}$ inches diameter. (See figs. 5. and 7., a side view and plan of one of the chains.)

There are four chains on each side of the bridge, each chain consisting of two rods; there are, therefore, eight rods on each side of the platform, to support it; and the total section of iron in the main chains is 2·6 square inches + 16 = 41·6 square inches.

The backstays are attached, by strong bolt-pins, 4

inches diameter, to cast-iron frames (see fig. 3.) fixed down on the abutments, at 95¼ feet horizontal distance from the points of suspension ; and thence they are continued obliquely downwards through the masonry of the abutments.*

The backstays are connected to these cast-iron frames by long open links, instead of round rods with eyes: there are sixteen links on each side, 3 feet 7 inches long, centre and centre.   The metal is 1⅝ deep by $\frac{13}{16}$ broad, and they widen at one end, to allow the large holding bolt to be put through them.   The vertical suspending rods are 1⅛ diameter, and are hung to the main chains by vertical links (see fig. 6.) 8½ inches long, 1⅝ broad by $\frac{11}{16}$ thick, which are jointed to the coupling plates of the main-chains by bolt-pins 1⅛ diameter.

118. The abutments (see fig. 3.) are built of masonry, 32 feet broad, and 38 feet from the front face, next the water, to the wall of the quay, up to which they are built, so as to form one mass with it.   Through each there is an arched opening for the path, along the side of the river, 13 feet wide ; the height from the ground to the top of the abutment is 14 feet.

119. The platform is laid upon cross joists, extending across in pairs, the same distance apart as the suspending rods ; each joist is about 5 inches broad by 12 inches deep ; the suspending rods go down through the space between the two, and are screwed up underneath with nuts, against strong cast iron washer plates.

On the cross joists are laid two lines of longitudinal beams, about 7 inches deep by 6 wide, at

* The author has been unable to ascertain how the extreme ends of the chains are fastened.   It is presumable they are held by cross-bolts behind the abutments.

each side of the carriage-way (see fig. 4.); and underneath each of these beams another longitudinal line of beams is placed, bolted to the upper one by bolts passing up between the cross joists. The space between the longitudinal beams, for the carriage-road, is filled up by a course of 4-inch longitudinal planking, arched upwards to the curve of the section of a road, and the space between them and the cross joists is filled up by cross pieces laid on them. A longitudinal beam is bolted also to the cross joists above them, outside of each footpath; and the platform is further stiffened by diagonal beams, extending underneath it from joist to joist.

For the footpath, a course of 2-inch planking, is laid crosswise over the longitudinal beams that rest on the cross joists.

The clear width of the bridge between the parapets is 25 feet 8 inches; viz. the carriage-way is 17 feet 8 inches broad, and each footpath 4 feet within the parapet. The extreme breadth of the bridge, from end to end of the cross joists, is 30 feet 8 inches.

The parapet consists of a strong longitudinal rail of wood, $5\frac{3}{4}$ inches broad by $7\frac{1}{2}$ inches deep, supported upon hollow iron pillars (see figs. 3. and 4.), the same distance apart as the suspending rods. This rail extends between the two chains on each side of the bridge, so that the suspending rods go down, alternately, inside and outside of it, but they are not attached to the parapet. The iron columns are $4\frac{1}{2}$ inches diameter outside at the top, and $5\frac{1}{4}$ at bottom, with a square foot cast out of the bottom, to stand on the longitudinal joist that extends along the outer edge of the footpath; and a long bolt goes down through each column, with a countersunk head in the top rail, and passes between the cross joists, to which it is made fast by a nut screwed up against a cast iron plate laid under them.

The platform vibrates very considerably when a carriage passes along it, and even with a horseman, notwithstanding the apparent stiffness and solidity of its construction, because the deflection is too great.

PONT D'ARCOLE, AT PARIS.

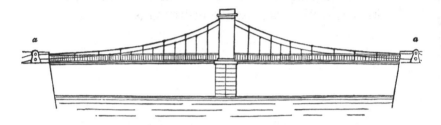

120. This bridge is erected across the Seine, at Paris, opposite the Hôtel de Ville. The chains are carried over a central pier built in the middle of the river (see the annexed sketch), and over two small iron frames, *a a*, fixed on the quay-walls, forming thus two semicatenaries, the chord of each of which is about 125 feet.

There are two chains on each side, each chain consisting of two sets of rods, 10 feet long and $1\frac{7}{8}$ inches in diameter, connected by coupling plates and bolt-pins. The total section of the chains is, therefore, 2·68 square inches + 8 = 21·44 square inches. The vertical suspending rods are 1 inch square: at the upper end they spread into a fork, which embraces the middle coupling plate between the two rods of each chain.

The parapet is of cast-iron, composed of upright rods 1 inch square, united by a top and bottom horizontal rail, and intermediate vertical round rods $\frac{3}{4}$ inch diameter.

121. The central pier is of stone, about 20 feet high from the platform to the cavities for the chains; 25 feet broad at the level of the road, and 10 feet

thick; with an arched opening through it 9 feet wide.

The holding bolts, which retain the chains to the short iron frames *a a*, are 6 inches in diameter.

The deflection, and consequently the vibration, of the chains, are very great.

BRIDGES MADE OF WIRE CABLES.

122. Messrs. Seguin, of Annonay, were the first to
introduce wire suspension bridges on the Continent.
They began by constructing a model of 62⅓ feet
span, and 2 feet wide. The platform was supported
by six cables, made of eight wires, each ·047 inches
diameter.

123. In 1823, they proposed to the French government a design for a large bridge of two openings, each of 278 feet, at Tournon, over the Rhône ;
and, in order to obtain experimental data for their
work, they erected a small iron-wire bridge, at Saint
Vallier over the river Galore. (See the annexed
sketch.)

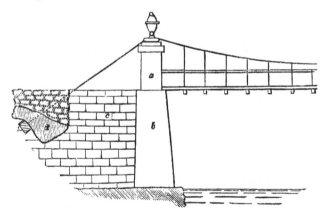

The span (centre and centre of the piers) of the
bridge of St. Vallier is 98·4 feet. The breadth 5½
feet; deflection 7¼ feet. The platform is 16·40
feet above high water. The supporting cables pass
over suspension pillars a 7¼ feet high, and 9½ inches

square ; and their ends are linked to iron rods, which are fastened to oak beams laid across in the masonry of the abutments, at the back of a large stone, *d.* The piers, *b,* are 16·40 feet high. The breadth at the level of the platform 10·49 feet; thickness at bottom 4·92 feet; and at top 3·28 feet. They join up at the sides to wing-walls, *c,* 3·28 feet thick at the base, and 1·64 feet thick at the top.

This bridge was loaded with 2500 kil. of gravel, and 70 persons = 4200 kil. = 6700 kil. = about 7 tons, without injury.

PONT DE TOURNON, CONSTRUCTED BY MESSRS. SEGUIN,
OF ANNONAY.

124. The bridge of Tournon, across the Rhône, between Tain and Tournon, was the first large wire suspension bridge erected in France.

The design was given in to the Conseil des Ponts et Chaussées in September, 1823, and accepted.*

The works were begun early in 1824, and the bridge opened in August, 1825.

The annexed figure is a sketch of the *Pont de Tournon*. It consists of two openings, each 278¾ feet wide.

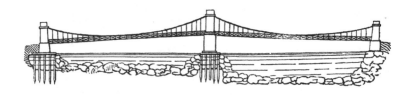

125. The abutments are 29½ feet broad by 23 feet thick, at low-water line. The opening through them for the road is 13 feet 1½ inch wide by 19 feet 8 inches high.

The central pier is 36 feet 9 inches wide by 16 feet

---

* In France, the roads and bridges are under the superintendence of the government engineers; and designs are referred to a board, composed of those of the highest rank, before they can be carried into execution.

5 inches thick at the level of the roadway, with starlings projecting out on each side, and rising 25 feet 7 inches above low water. The opening through it, 20 feet 4 inches wide.

126. The platform is 13 feet 9 inches wide between the parapets; but it begins to widen at 59 feet from each side of the central pier, and spreads out to the width of the opening through that pier. It is divided into three parts, viz. a central carriage-way, 7 feet 7 inches wide, and two footpaths. It is composed of oak cross bearers, 16 feet 5 inches long, $11\frac{3}{4}$ inches deep by $6\frac{1}{4}$ inches broad, placed $3\frac{1}{4}$ feet apart centre and centre. It is $25\frac{1}{2}$ feet above low water at the middle, and 18 feet $7\frac{1}{2}$ inches at the ends.

For the carriage-way a course of 3-inch planking is spiked down to the cross beams; and then over that a course of cross planks of poplar, $2\frac{3}{8}$ inches thick.

The footpaths are raised above the carriage-road on the two lines of longitudinal fir-beams, $5\frac{3}{4}$ inches by 11 inches, bolted on the cross beams, and covered with oak planks $2\frac{3}{8}$ inches thick.

The parapet is 4 feet above the footpath, formed of oak rails 4 inches by $9\frac{3}{4}$ inches, and diagonals 4 inches square.

127. The main cables are twelve in number, six on each side of the bridge. Each cable contains 112 wires of No. 18. viz. (3 millim. diameter = ·0394 × 3 =) ·1182 inches diameter. They are put together in three lengths, each of 98 feet 5 inches. The ends are made into loops, and coupled by 56 turns of wire passed round two cast-iron sockets, or half pullies, fitted into the loops. The cables are placed

side by side on the summits of the abutments; the
two inner cables being there 17 feet 4½ inches apart;
and the outer ones 21 feet 3¾ inches apart.   Hence,
as the platform is only 13 feet 9 inches wide, the suspending wires are not quite vertical.

The cables are carried down vertically from the
summits of the abutments close behind them at the
Tournon end of the bridge.   They are fastened to
long iron open links, made of bar iron, 1·18 inch
by 2 inches.   The lower ends of each set of links are
retained by horizontal keys, held down to the rock at
one end by iron cramps, and at the other, by blocks
of granite laid in the foundation.

On the summits of the abutments, the cables are
fastened to a strong bolt, put through a cast-iron
saddle, 4 feet long, 1 foot 7½ inches broad by 1 foot
7½ inches thick, which is capable of rocking upon its
bearings.

At the *Tain* end there was no rock.   A gallery
of masonry, therefore, was formed, 4 feet 11 inches,
high, covered by blocks of stone, with tunnels through
them for the bars which were retained under the roof
of the gallery by cross keys.

On the summit of the central pier the cables pass
over a semicircular bed of masonry, extending all the
breadth of the pier, and covered with sheet-iron.
They descend thence to within 16 feet 5 inches of
the lower course, and are there fastened to iron bars
1·18 inch by 2 inches, which are keyed into the lower
course.

128. The following is M. Seguin's statement of
the strength of the parts, and the strains they are
exposed to : —

	Tons.
Weight of Platform - - -	69
Strain that it brings on the abutments	98·6
Resistance of the central pier - -	197·2
Ditto, or breaking strain of cables -	443·7
The vertical cables would bear -	3549·6
The abutments would resist a strain of	591·6

The bridge was tried before being opened, and
bore, without injury, in dead weight, 65,900 kil.
on one opening $= 65$ tons; and afterwards 61,250
kil. $= 60\frac{1}{2}$ tons in dead weight, and 7900 $= 7\cdot8$ tons
in people.*

The total load, therefore, was $60\frac{1}{2} + 7\cdot8 = 68\cdot3$
tons + the weight of the bridge 69 tons $=$ in all
137·3 tons. And the deflection being $\frac{1}{11}$th of the
chord-line, the tension is about 1·46 times the weight.
Hence the strain on the cables and on the abutments
is $137\cdot3 \times 1\cdot46 = 200$ tons, or $\frac{1}{2\cdot22}$th part of the
breaking strain of the cables; and a little more than
$\frac{1}{3}$ of the resistance stated for the abutments.

This bridge is 13·75 feet broad by 278·75 feet
long $= 3832$ square feet; and could contain, therefore,

---

* This was the proof load required by the government. The
standard proof for suspension bridges in France being 200 *kilo-*
grammes per square *mètre*, which amounts to about 41 lbs. per
square foot. In this country, bridges are generally expected to
carry a greater load than that. The standard adopted in the
calculations throughout the work for the load that a large bridge
may have to bear, is at the rate of 70 lbs. per square foot of the
platform, viz. *one person for every two square feet of surface*,
averaged at 140 lbs. weight per head. It is presumable, that, in
France, the bridges being more immediately under the control
and inspection of the government, they will not permit them to
be loaded beyond their own standard. Therefore, in judging of
the strength of bridges in that country, we must consider it in
reference to the load they are intended for.

$\frac{3832}{2}$ = 1916 persons, which × 140 lbs. = 26,824 lbs. = 119·7 tons.

The entire strain might therefore be, in dead weight, 119·7 tons + 69 = 188·7 × 1·46 = 275·5 tons, or $\frac{1}{1\cdot61}$ of the breaking strain of the cables, and $\frac{1}{2\cdot15}$ of the stated resistance of the abutment.

SUSPENSION BRIDGE, MADE OF IRON WIRE, AT GENEVA.

Ground Plan of the Piers.

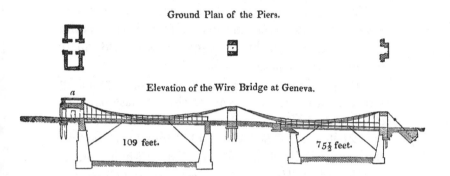

Elevation of the Wire Bridge at Geneva.

109 feet.   75½ feet.

129. The bridge (shown in the figure above)
crosses the double ditch of the fortifications at Geneva. It is made entirely of small wires laid together
side by side, and bound up into small cables, by wire
wound spirally round them. It was erected under
the direction of Col. Dufour, an officer of engineers.
The estimated expense was 16,000 francs (640*l.*), and
the real cost a very little more.

The entire distance between the bastion of the
inner wall at *a*, called *le Bastion du Pin*, and the
opposite bank of the outer ditch, is 268¾ feet. The
inner ditch is 109 feet wide, the outer one is 75½ feet
wide, and the embankment or counterguard between
the two is 82½ feet wide. The breadth of each opening is 131 feet clear between the piers. The platform
of the outer opening is 91 feet long, although the
ditch is only 75½ feet wide, because the bank of the
ditch slopes. See the figure. The piers are built of
masonry. They are 13·12 feet high from the ground,

and 12½ feet broad, with openings through them 6 feet wide and 10 feet high.

The foundations are laid upon piles about 12 feet long. The platform of each opening is supported by six cables, each containing 90 wires of No. 14. wire, viz. 2·10 millimètres = 3·464 mill. sectional area, or, in English measures, ·082 inches diameter, and ·00528 inches sectional area. Hence the entire section of wire in the bridge is 90 × 6 × ·00528 square inches, = 2·85 square inches.

Colonel Dufour found the breaking strain of No. 14. wire 209 kilo. ; whence the ultimate strength of the six cables is 209 × 6 × 90 = 112,860 kilo. × 2·21 lbs. = 249,420 lbs. = 111½ tons.* If we take the standard, deduced from Mr. Barlow's experiments (*vide antè*, p. 16.) 38½ tons per square inch for the ultimate strength of iron-wire, we have 2·85 square inch. × 38·5 = 109¾ tons. The accordance of these two results, obtained by separate and independent experiments, is a confirmation of the correctness of each.

130. The main cables pass over the pier in grooves formed in bed-stones laid on the tops of the piers, 3¼ feet long, 2·624 feet wide, and 2¾ feet thick ; the edges of the grooves are rounded off where the cables come on them, and are covered with a thin brass plate. The ends of the cables at the town end of the bridge are fastened to vertical bars 7½ feet long, and 1⅜ inches square, which descend close against the back of the pier ; their lower ends are linked to horizontal bars 2 inches deep by 1·36 inch broad, and 3¼ feet long, laid on edge in the foundations of the pier ; the backstays being thus carried down perpendicularly, this pier has to bear all the drag of the bridge to pull it over : and is made larger and stronger

* One kilogramme = 2·21 lbs. avoirdupois.

than the other in proportion. (See the figure.) At the opposite end of the bridge, the cables are fastened to inclined bars also 1⅜ inches square.

The cables are in several lengths, viz. : there is one long cable extending across each opening, and three short ones over each pier. The latter may, in fact, be called adjusting links, inasmuch as by varying their lengths the main cables were adjusted ; for the three cables on each side of the bridge are placed side by side on the tops of the piers, in order to avoid the complication and expense of having a saddle for each cable. But their position is altered on quitting the piers, viz. : they are brought nearer together sideways, and one under the other, as they approach the middle. For that, it was necessary for the cables to be of different lengths, which were adjusted by varying the different lengths of the short connecting cables that lie on the tops of the piers. The ends of the long cables are fastened to the short ones by links made of wires (see the figure below); loops at the ends of the cables are made like the loop of a cord, by simply turning the wires back upon themselves, and then binding the loose ends and the cable together by a wire wound tight round them at *a a*. (See the sketch.)

Plan of linking the Cables together.

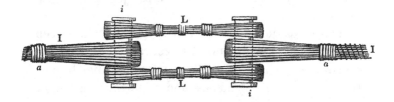

The cables I and links L are united by bolt-pins,

made hollow, that they may be of large diameter
without being heavy.

N. B. — Col. Dufour tried various ways of fastening the cables
together, and ultimately selected this as the simplest, and, at the
same time, sufficient in strength, the cables breaking before the
fastenings would untie. He found that, with No. 14. wires, thus
bent into a loop, and bound round with annealed wire of No. 4.
strength (viz.: 0·0335 inch. diam.) over 2 inches of the length
of the cable, the fastening did not yield, even when it was covered
with oil.

Where the cables are looped round the ends of
the iron bars for the back-stays *b*, a small semicircular
brass washer, *a*, is interposed to prevent the wire
being bent too abruptly. (See the sketch annexed.)

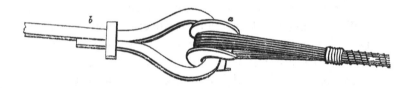

131. The platform consists of cross joists, suspended from the main cables by short vertical cables,
each containing 12 wires. The upper ends are
looped round the main cables, and the ends tied
round by a wire in the way described of the loops at
the ends of the cables, and the lower ends are simply
looped round the ends of the cross joists, and also
tied.

The cross joists are 7½ feet long, 5½ inches deep
by 4½ inches broad, and 4¼ feet apart centre and
centre. On them are laid five longitudinal beams,
6 inches deep by 4½ inches, and over those a course
of 1½-inch planking.

The parapet is 3½ feet high, composed of upright

rods of iron $\frac{3}{4}$-inch diameter, with a top rail $\frac{3}{4}$-inch diameter, and two other rails of flat iron $\frac{3}{4}$ of an inch by a $\frac{1}{4}$ of an inch. The parapet is inside of the suspension wires, to guard them against injury.

The platform is steadied beneath by diagonal ties, about 25 feet from each end, one end fastened to the platform, and the other to eye-bolts let into the walls of the ditch.

132. The load on the larger opening is thus calculated by Colonel Dufour: —

	Kilo.		Lbs.
160 Persons - -	10,500	=	23,205
Weight of bridge, platform, &c. - -	7200	=	15,912
Chains, &c. - -	700	=	1547
Chance Load - -	100	=	221
	18,500	=	40,885 = $18\frac{1}{4}$ tons.

The angle of direction of the chains is 19°, and the tension for that angle is about 1·5 times the weight $= 27\frac{3}{8}$ tons.

The ultimate strength of the cables is, as we have seen, $111\frac{1}{2}$ tons, and they might be safely loaded with one third of that, or 37 tons, instead of $27\frac{3}{8}$.

N. B.—The platform of this opening is 110 feet long by $6\frac{1}{2}$ feet broad = 715 square feet. Taking, therefore, our standard of 70 lbs. per square foot, it might happen to be loaded with 50,050 lbs. = $22\frac{1}{3}$ tons; the weight of the platform and chains, &c., being a little less than eight tons. The total weight suspended would then be $30\frac{1}{4}$ and the tension $45\frac{3}{8}$ tons.

This bridge vibrates very much, from the great slackness and lightness of the construction. It is well adapted for its purpose, being intended only for foot passengers, but it would be dangerous to allow it to be completely loaded, or to allow more than 50 men to march over it in step.

SUSPENSION BRIDGE AT ARGENTAT OVER THE RIVER
DORDOGNE, ERECTED BY MONS. L. T. VICAT, INGÉNIEUR
EN CHEF.

Longitudinal Section.

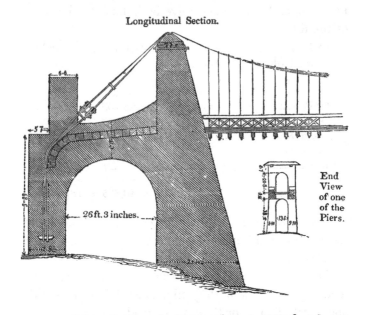

End
View
of one
of the
Piers.

133. The platform is (100 mètres) 328 feet long,
and 13¾ feet wide ; *viz.* the carriage-road is 7¾ wide,
whit a footpath of 3 feet on each side.

It is curved upwards at the rate of 1 inch for
every 100 inches. It consists of 101 cross joists
3·28 feet apart centre and centre, 7⅛ inches by 9½
inches, and 16 feet long. On those are laid four
courses of beams on edge, 3¼ inches broad by 8¼
deep, forming the boundaries of the carriage-road
and of the two footpaths.

The parapet consists of two horizontal rails of
wood, 3¼ inches broad by 8½, united by diagonal

braces, and bolted to the wood-work of the platform by ¾-inch bolts.

The framing of the platform is covered with one course of 1½-inch planking, for the footpaths, laid on the longitudinal beams. For the carriage-way by a course of 3-inch planks, laid on the cross joists, ⅜ths of an inch apart; and over that a second course of 1½-inch poplar-wood planking, nailed crosswise over the longitudinal planking.

134. The platform is suspended from the main cables, by 101 vertical suspending cables on each side, composed each of 40 wires of No. 18. strength, *viz.* of 0·11808 inches diameter. Their upper ends are attached to 6 cables on each side of the bridge. The chord-line of the cables is 350 feet 5 inches. On the summits of the piers they lie side by side; but, as they approach the middle, they are brought one under the other, so that they have all different deflections. The deflections are as follows:—

			Feet.
Of the upper one,	-	-	22
second	-	-	23·42
third	-	-	24·21
fourth	-	-	24·98
fifth	-	-	25·75
sixth	-	-	26·5

The number of wires composing each cable, is according to its deflection; *viz.*—

The upper one contains		216
second	-	- 210
third	-	- 202
fourth	-	- 196
fifth	-	- 190
sixth	-	- 186

The cables pass away towards the fastenings from the summits of the suspension pillars, at an angle of 45°, and, just before entering the masonry, are linked with short links and bolt-pins to long curved iron links, which lie against an arch of masonry (see the figure), and are fastened to another set of long links, the ends of which embrace a cast-iron anchor (see the figure) let into the solid of the masonry.

The main piers are of the shape and dimensions shown by the drawing (figs. 1. and 2.). The whole of the masonry is composed of *schistus* of middling hardness, cemented with mortar made of hydraulic lime and river sand. The only part where blocks of stone are used is where the chains bear against the masonry in the abutments. (See the figure, p. 120.)

Cable.

Fig. 1.

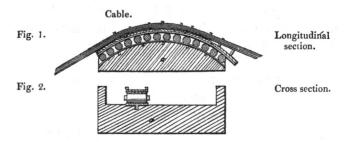

Longitudinal section.

Fig. 2.

Cross section.

Each of the cables, at the points of suspension, lies upon a sort of carriage of rollers (see the figure above), *viz.* a cast-iron bed, *a*, with an arched upper surface, is bedded in the top of the pyramid; on this are riveted, side by side, six iron plates; on which are laid, under each line of cables, fourteen small cast-iron rollers (5 centi-mètres =) 1·970 inches diameter, $2\frac{1}{10}$ inches apart, centre and centre, kept apart by iron plates fitted on to their necks. Over the rollers are laid thin iron plates, and on these the cables lie. (See fig. 1.) Hence the cables can move a little endwise, *viz.* about 4 inches over the tops of the pyramids, without injuring the masonry.

135. Before being opened, this bridge was proved, according to the government standard of 200 kilo. per square mètre, or 41 lbs. per square foot, by loading it with sand in three successive loads, in all 84,000 kils. =   -   -   -   - tons 83
the platform itself, and cables &c. are estimated at 105

Total 188 tons.

The first load was about 21 tons, and produced a deflection of $1\frac{3}{8}$ inches. The second made up 41 tons, and produced an additional deflection of $1\frac{5}{8}$ inches, and with the whole load the platform had sunk $7\frac{1}{4}$ inches below its original level.

The platform rose as the load was removed, and when the load was all taken off, had risen to within $2\frac{1}{4}$ inches of its proper level ; and the next day it had risen to within $1\frac{3}{10}$ inch.

The deflections of the main cables being different, their tensions are consequently different. To calculate, however, for example, the strain on the upper pair of cables. Their deflection is $\frac{350.5}{22} = \frac{1}{15.93}$th of the chord-line. The tension at the points of suspension, with that deflection, is 2·041 times the weight : and as the two upper lines of cables have to bear only $\frac{1}{6}$th of the whole weight, the strain on them is $\frac{188}{6} = 31\cdot3$ which × 2·041 = 63·88 tons, or nearly 32 tons on each of those cables. Now each contains 216 wires of No. 18. *viz.* ·1181 inches diameter ; or (·011 square inches × 216) = 2·376 square inches of iron wire in each cable.

Mons. Vicât states the ultimate strength of the

wire used, from experiment, 1165 lbs., which is at the rate of 47 tons per square inch. Hence the ultimate strength of the cable would be $(2 \cdot 376 \times 47 =)$ $111 \cdot 67$ tons, and the stretching strain one third of that, or $37 \cdot 2$ tons.

The cables will therefore bear safely more than their proof load.*

These trials were made in September, 1829, and the bridge was opened in the latter end of that month.

---

* It would be very desirable to try, experimentally, the ultimate strength of a *cable of wires*, and to compare the result with the ultimate strength of the wires themselves ; for it is much to be doubted whether a cable possesses the strength that is computed for it by *taking the number of wires and multiplying it by the strength of each individual wire.*

BRIDGE AT JARNAC, CONSTRUCTED BY MONS. J. P. QUENÔT.

136. The town of Jarnac is built on the right bank of the Charente, and has considerable trade with the town of Cognac, in distributing the famous Cognac brandies through the interior of France.

Before the erection of the suspension bridge the communication was kept up by a ferry, frequently dangerous, and sometimes impassable; and several attempts had been made by the government engineers to have a stone bridge erected, but the expense prevented their proposals from being acceded to.

A design for a suspension bridge of iron-wires was afterwards proposed by Mons. Quenôt, and adopted, and the excavations for the foundations of the main piers were commenced in May 1826.

The drawing below contains an elevation and plan of the Jarnac bridge.

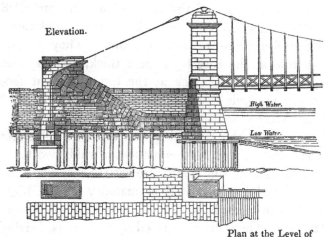

Elevation.

High Water.

Low Water.

Plan at the Level of
the Platform.

Under the bed of the river, and a great part of the banks, is a stratum of calcareous rock, disposed as shown by the drawing.

The whole of the masonry on the left bank is founded on piles. On the right bank only the main pier is built on piles; the remainder on a framework bedded on the rock.

137. The piers stand each on seventy-eight oak piles, driven into the ground 10·8 feet, about 1 foot diameter, and 4 feet apart, centre and centre.

The piles were cut off at 1·8 feet below low-water mark, and over their heads were bolted longitudinal and cross beams $30\frac{1}{4}$ feet long, and 1 foot square, making thus a framework with square openings about 3 feet square, which were filled up with rubble.

The piles for the back or wing walls are eighteen in number, extending along in two lines, each covering a space of $43\frac{3}{4}$ feet by $4\frac{1}{4}$ feet.

And the piles for the masonry to which the chains are fastened are in two rows, covering a space of $28\frac{1}{4}$ feet by $7\frac{1}{4}$ feet.

The basements of the suspension piers are 21·8 feet high above the framework of their foundations, and 20 feet above the low-water-line. They are $40\frac{1}{4}$ feet wide at the base by 18 feet thick, and $34\frac{1}{2}$ feet broad by $12\frac{1}{2}$ feet thick at the top, the front face sloping upwards from the water. (See the figure.) The external walls are built of close masonry, the interior of rubble, and they contain 3916 cubic feet of stonework, and 7356 cubic feet of rubblework.

The ends of the chains are fastened at each end of the bridge into a well of masonry 37 feet high. The mean breadth is $7\frac{1}{2}$ feet, and the mean length 18 feet. The wells are formed by four walls, leaving an opening $3\frac{1}{4}$ feet square. At the bottom is a stone

containing 77½ cubic feet, with a hole made through
it, for the chains to pass through, 3¼ feet by 4
inches. This stone is fastened down by strong iron
cramps to the framework on which all the masonry of
the abutments stands, and which contains 4787 cubic
feet. The curve against which the chains lie is 7¼
feet radius, beginning 21⅓ feet above low water, and
formed of two blocks of stone. In front of the wells
is built an oblique buttress, or wall of masonry (see
the elevation), as far as the base of the main piers.

The columns over which the cables pass are four in
number, viz. two on each pier, 30½ feet high ; and
are 8·87 feet thick in the direction of the line of
road, by 12 feet broad. They are built entirely of
blocks of stone, and contain each 2027 cubic feet of
stone.

138. The main cables are formed of wire, No. 18.
viz. ·1181 inches diameter, and 0·011 inches sectional area.

Before using the wire it was left for an hour in
a cauldron of boiling linseed oil, thickened with litharge and a few ounces of soot. When taken out
it was hung up in bundles, and dried in three or four
days.

Each cable contains 300 wires, laid side by side,
and bound together at intervals of 3¼ feet by a binding of smaller wire wound round about 4 inches length
of the cable.

The cables were made by four men, carrying a reel
full of wire backwards and forwards between two
fixed posts, stretching and fastening one wire at a
time ; and so laying 300 together parallel to each
other, taking care not to twist any, and that they
should all be equally deflected from a straight line. As
fast as the wires were stretched and straightened, a

man followed, covering them with varnish made of 16 parts of turpentine, 8 parts of litharge and linseed oil, and 16 parts of resin.

139. The platform (see the following sketch) is borne upon cross beams. The difficulty of procuring good wood of sufficient dimensions prevented their being of one piece ; they were therefore framed (as shown by the figure) of separate pieces of oak bound together by iron straps, covered over with pitch at the parts in contact. These framed cross bearers are 2·13 feet deep by 1 foot, and their length is 28¼ feet.

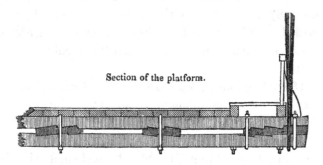

Section of the platform.

On them are bolted four lines of oak beams, 9½ inches by 7¾ inches, and about 16 to 19 feet long. Between these, for the carriage-way, is a course of 4¾-inch oak planking ; and over that, to protect it, a course of cross planking of poplar 2¼ inches thick, a thick layer of pitch being interposed between the two.

The upper rail of the parapet is 7 inches by 3¼, with vertical standards and diagonals 3¼ by 2½.

140. The chains pass over cast-iron supports placed on the tops of the columns. These are a sort of grooved sectors (see the following figures), fitted each on a semi-cylindrical bar, cast out of a bed plate 3½ feet by 1·312 feet at the base, and 3¼ inches deep in the middle. The supports can thus move backwards and

forwards a little on their semicylindrical axes, and
allow for a little play in the cables.

Edge view.   Front view.

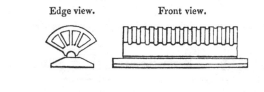

Plan of the cast-
iron plate.

Plate.

Cables.

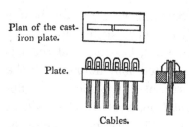

The cables were each
made in six lengths,
or links. The extremi-
ties were passed through
two holes 18 inches
long by 3¼ inches wide,
in cast-iron plates 4 feet long by 2 feet wide, and
9½ inches thick (see the figures); one such plate
being placed in each of the wells under the large
stone. The end of each cable is looped round a half
pulley of iron, through which is put a cross key bearing against the holding plate. The opposite end of
the cable was raised by men placed on the top of the
suspension column, by means of a rope and windlass, up to *a* (see the elevation, p. 125.), and there
united by iron keys to the next piece lying over the
suspension columns.

The third piece of cable reached thence to the middle of the bridge, and the two loose ends were united in
the middle by means of iron clamms (see the figure
below), which laid hold of them ; and were brought
together by a pair of screws, *a*, until the cables were
strained to the proper tension. The loop ends of the
cables were then bound together by as many turns of
wire as the main cables themselves contained. A
scaffolding 6½ feet square was erected in the middle

K

of the river for the men to work upon in these operations. The twelve cables were got up in fifteen days.

Side view of the clamms for bringing the cables together.

141. On the 2d of September 1827, the bridge being completed, was tried before being opened, by loading it uniformly with sand, beginning at the extremities. The loading lasted four hours. As the weight was gradually increased, the platform was depressed; and, with a total load of 106768 kil. = 105 tons, sunk 3¼ inches. The load being removed, the platform had risen 2¼ inches; and, in about fourteen hours more, recovered its original position. No injury whatever was done to the bridge by this load.

Calculation of the Strength of the Bridge.

142. M. Quenôt estimates the suspended weight of the bridge itself, cables, keys, &c., and wood-work - 148,232 kil. =

	Tons.
and wood-work 148,232 kil. =	147
Load brought on it	105
Total	252

The chord-line is 229½ feet, and the deflection 19·68 feet, or $\frac{1}{11\cdot66}$th of the chord line. The tension at the points of suspension for that deflection is about 1·52 times the weight, and the strain on the cables is therefore (252 × 1·52 =) 383 tons.

The main cables contain in all (300 wires × 12

cables × 0·011 square inches =) 39½ square inches of iron.

Taking the strength of the wires at the standard of 38½ tons per square inch, the cables would bear, without injury, (39 × 12·8 =) 505½ tons, or 122½ tons more than the strain that the bridge underwent at the trial. And if, as may be presumed, the wire was as strong as the No. 18. wire used in the Argentat Bridge (*vide ante*, p. 124.), the cables would bear (39½ × 15·6 =) 616 tons of strain.

STEEL SUSPENSION BRIDGE AT VIENNA, OVER THE
DANUBE.

143. The following account of this bridge is
taken from a description published at Vienna, by
Herr Ignaz Edlen von Mitis, who designed and
erected it.

It was begun in June, 1827, and opened June,
1828.

The span, or horizontal distance between the points
of suspension, is 334 English feet; and the middle
deflection 21·4 feet, or $\frac{1}{15\cdot6}$th of the chord-line.

The length of the curved chain is stated 339·17
feet.

The suspension piers are of masonry, both alike,
40½ feet by 21·4 feet.

The platform is 324·21 feet long, and 11 feet 10½
inches wide.

144. It is supported by two chains, each composed of four lines of steel bars in links of the same
shape as in the Menai Bridge, 6·42 feet long, from
centre to centre of the eyes, and 2·42 inches broad
by 0·8 inches thick; united by steel coupling plates,
and bolts 2·4 inches diameter. The section of each
of the bars of the chains is therefore 1·936 inches;
and the total section of steel 1·936 × 8 = 15½ square
inches.

The estimated weight of the chains, Lbs. Avoir.

coupling plates,and bolts is -	19,836
Vertical rods - - - -	2534
Beams of wood for platform - -	13,705
Planking - - - -	41,170
Parapet, &c. - - - -	10,358

Total weight of the bridge * 87,603

145. H. Mitis computes the load that may come on the bridge at the rate of about 15 men per square fathom, of 115 Vienna pounds weight each. The platform contains 93 square fathoms, —
Hence the whole weight is 160,425 Vienna Tons.
pounds $=$ 176,788 lbs. avoir. or $=$ - 79
The weight of the bridge itself is 87,603 lbs. $=$ 39

Total 118

The angle of direction of the chains is about 14° 25′, the sine of which is 0·24897. Whence the tension at the points of suspension $= \frac{118}{2 \sin. c} = \frac{118}{\cdot49794} = 237$ tons.

146. Now, as to the strength of the chains, Herr Mitis states that he found *the strain that would stretch steel bars, by an average of thirteen experiments,* $=$ *47,125 lbs. or 21 tons per square inch.* The bars of his bridge were proved to 25¾ tons per square inch, without being stretched thereby ; and the coup-

* The measures and weights in H. Mitis' description, are given in Vienna measures, which have been converted into the corresponding English ones.
The Vienna foot = 1·07 feet English.
The Vienna pound = 1·102 lbs. English avoirdupois.
The Vienna fathom = 6 Vienna feet (or 6·42 feet English).

ling plates to 21½ tons per square inch. Hence all the bars of the main chains will bear $15\cdot5 \times 25\cdot75 =$ 399 tons without stretching, or 162 tons more than they will be loaded with. And as the coupling plates contain $19\cdot37$ square inches, they will bear $19\cdot37 \times 21\cdot5 = 416$ tons, or 219 tons more than they will be loaded with.

147. H. Mitis' work contains a great many tables of experiments on steel, &c., and investigations on various points connected with suspension bridge buildings; and he appears to advocate the superiority of steel over iron as a material for suspension bridges. In this country the price would be a prohibition, even if it had all the advantages possible. But its chief apparent superiority, viz. *strength as compared to weight,* is not really an advantage, if it be used as a material for the chains of suspension bridges, because weight and resistance to motion are, to a certain extent, desirable (as stated, *antè,* p. 27.). For the vibration produced on a very light but strong bridge, by a *moving load* greatly beyond its own weight, is far more dangerous to its stability, than an increase of its own inherent weight would be. In this steel bridge, for example, the weight of the chains is only 8 tons 1910 lbs; and the total weight of the bridge is 39 tons. Now, the load estimated by H. Mitis is 79 tons, or above twice the weight of the bridge, and about 8¾ times the weight of the chains: And if we were to compute the possible load by the standard adopted in this work, *viz.* 70 *lbs., or* $0\cdot03125$ *tons per square foot,* the platform of this bridge containing 3855 square feet, might be loaded with 120 tons, or more than 3 times its own weight, and more than 13 times the weight of its chains.

The practical effect of such an excess of lightness

is, in fact, felt in this bridge; for H. Mitis says, himself, that from that cause the bridge vibrates and swings during heavy winds, or under moving loads, more than he could desire; and he admits it to be a drawback on the use of steel.

When, therefore, from any special reasons, it is desirable to make a bridge very light in proportion to its strength, the chains ought to be tied together crosswise, as in the Menai Bridge, and steadied by under-ties; and the parapets should be made very stiff, if they cannot be made heavy. In fact, tendency to vibration *must* be counteracted; and if it cannot be counteracted by downright mass, it must be done by mechanical arrangement.

SECTION V.

DESIGNS, AND BRIDGES IN EXECUTION.

BRIDGE PROPOSED OVER THE SOUTH ESK, AT MONTROSE,
BY G. BUCHANAN, CIVIL ENGINEER. 1823.

148. A design was proposed, in 1823, by Mr. Buchanan, for a suspension bridge over the river South Esk, at Montrose, which, although it has not been executed, possesses some peculiarities that render it worthy of notice.

The following are some particulars relating to it, taken from a report of Mr. Buchanan's, printed in the Edinburgh Phil. Journ. Nos. XXI. and XXII. 1824.

Span 420 feet. Deflection of the chains 60 feet, or ¼th of the opening. Height of the suspension pillars 60 feet above the roadway.*

149. *Main chains* to consist of thirty-six single chains on each side of the bridge; each chain formed of round rods 15 feet long and 1½ inches diameter, with

---

* Mr. Buchanan recommends this proportion for the deflexion in preference to a greater tension, as producing a diminution of expense, consequent upon the diminution of strain on the chains. We have no examples of large bridges built with so great a slackness of the chains. It draws with it a greater liability to vibration, and an increase in the height of the suspension towers. These things considered, it may be doubted whether the inconveniences of so great a deflection, would not overbalance its advantages. The vibration of the two bridges at Paris, described in the preceding section, is very striking, and may, in part, at any rate, be attributed to the great deflection of their chains.

a shoulder, 2 inches diameter and 1½ inches long, formed at each end. The rods to be coupled by cast-iron sockets, about 1¼ feet long and 4 inches diameter, cast in two halves, or semicylinders, to fit the bars and catch against the shoulders, and the two semicylindrical sockets to be bound together by wrought-iron hoops.

The thirty-six bars of each chain to be ranged so as to make a square of six single chains each way, forming a square of about 16 inches to the side, the couplings being about 5 feet apart on the length of the chains.

The section of each chain would be 63½ square inches; consequently, the entire section of the supporting chains 127 square inches. The section of the metal of the coupling sockets to be about five times that of the rods.

150. *The suspension pillars.* Two pyramids at each end of the bridge, 50 feet high, 9 feet square at the bottom, and 4 feet square at the top, to stand on a pedestal 11 feet square and 11 feet high, the bottom of which would be level with the roadway. The pedestal to stand on a stone pier 12 feet square at the top, and 20 feet square at the bottom, which would be on a level with low-water mark. The pyramids to be made of iron plates ½ an inch thick, with flanges, bolted and rusted together. The hollow space enclosed between the plates of the pyramids to be filled with stone.*

The pedestal to be of stone; the base or pier of

---

* A similar plan has been pursued in the erection of the colossal statue of San Carlo di Borromeo, at Arona, in Piedmont. The statue is between 60 and 70 feet high, and is made of copper plates riveted together round an internal column or core of masonry, to which the plates are fastened by iron-stanchions.

ashlar. The pyramid to stand on an iron plate laid
on the top of the pedestal; the pedestal to stand also
on an iron plate laid on the pier; and the pier to
stand on a platform laid on piles driven hard into the
ground; and one long chain to run right down
through the whole, to bind them together in one
mass.

151. *Fastenings of the ends of the chains.* Each
rod to be fastened to a cast-iron bracket, bolted to the
top of a wooden platform 20 feet square, attached to
a series of piles driven hard into the ground, the fastenings to be about 20 feet below the ground.

The thirty-six bars of each chain to spread out as
they approach the platform from their compact square,
so as to cover the platform of 20 feet square, and then
each rod to be attached to its bracket. The platform to be loaded with sand or gravel, to the amount
of 500 to 1000 tons for each end of the bridge, viz.
to be equal to the greatest load the bridge would
have to bear.

152. The load is estimated by Mr. Buchanan as
follows : —

	Tons.
Computed load of iron in roadway, vertical rod, &c.	180
Gravel	220
Chains	100
And 7000 people, which it is assumed the bridge would bear safely, reckoned each at 150 lbs.	470
Total	970

The tension of that load at the points of suspension, with a deflection of ¼th, would be a little less

than 970 tons, *viz.* 970 × ·99 = 960 tons. And therefore the chains would be strained at the rate $\frac{960}{127}$ = about $7\frac{1}{2}$ tons per square inch.

153. *The roadway* to be 30 feet wide, clear of the chains and side rails ; two foot paths, each 4 feet wide, to be borne by cross joints of malleable iron, 5 feet apart, $3\frac{1}{2}$ feet deep in the middle, and well trussed, and also strengthened by longitudinal plates of malleable iron 2 feet deep. On the cross bearers, cast-iron plates $\frac{1}{2}$ an inch thick to be laid, and on those, gravel or road metal 4 or 6 inches deep.

The vertical suspending rods, two $\frac{3}{4}$-inch rods on each side of the bridge, embracing the chains at the couplings, 5 feet apart.

The height of the platform above high water is not named ; by Mr. Buchanan's drawing it would be about 20 feet.

And to allow ships to pass the bridge, a draw-bridge was proposed at one end, *viz.* The platform at that part to be made very light, and to open, and be lifted up about an axis in the pier, being drawn up by a rope passing over a pulley in each pyramid, wound round a barrel within the pedestal, and turned by a winch applied on the axis. The side rails or parapets, and the vertical rods, to be made at that part in two pieces, so as to fold up inwards on a hinge joint, to-wards the centre of the bridge. To stiffen the bridge, the vertical suspending rods, both in the central open-ing, and on the land sides of the pyramids, were to be tied together by a lacing, or sort of net-work, of wrought-iron rods.

DESIGNS FOR A SUSPENSION BRIDGE OVER THE AVON,
NEAR CLIFTON.

154. The position selected for a suspension bridge
near Clifton is singularly adapted for a structure of
that kind; for the river Avon is there narrow, and
enclosed between rocks ranging between 200 and 300
feet in height, and not 800 feet apart at the summits.
On one bank they rise up rapidly in wild rugged
masses; on the opposite they are less abrupt, and
clothed with foliage. It would have been exceedingly difficult, if not impracticable, in this situation, to
erect a stone or wooden bridge; while, on the other
hand, the success of the Menai, and other suspension
bridges of large span, justified the expectation that
one on that principle might be thrown across.

155. In September, 1829, the persons interested in
the erection of the bridge issued proposals for designs to be sent in, offering a premium for the best.
A considerable number were accordingly prepared,
among which one by Mr. J. Brunel was selected,
and adopted for execution in March, 1831.

156. The author is indebted to Mr. Brunel for
the following dimensions of his intended bridge : —

	Feet.
The horizontal distance between the points of suspension - - - 700	
The deflection of the chains - - - 70	
The breadth of the platform - - 33	

It is divided into a carriage-way 20 feet broad, and
two footpaths.

Height of the main towers above the platform, 80 feet. They are to be built of compact red sandstone, and are 52 feet broad, by 36 thick at the level of the platform. They thence rise up pyramidally. At the summit, the platform for the main chains to rest upon, is 30 feet broad by 20 feet.

157. The platform is to be supported by four main chains, each composed of sixteen bars, $7\frac{3}{4}$ inches broad by 1 inch thick ; 18 feet long from centre to centre of the eyes. There are to be no coupling links ; but the ends of the bars to be linked directly together by bolt-pins 5 inches in diameter.

The dimensions of the chains are diminished towards the middle of the opening, where they are $\frac{1}{13}$th less in section than at the points of suspension. The entire section of the chains at the strongest part is, therefore, $7\frac{3}{4} \times 64 = 469$ square inches ; and, at the weakest, 458 square inches.

The back stays are carried off at very nearly the same angle as the central chains ; and are to be carried down from the surface of the ground about 100 feet, into the solid rock, to their fastenings.

At the points of suspension, the chains are to pass over separate tables running on horizontal rollers.

The height of the platform above high-water mark is 240 feet.

	Lbs.
The estimated weight of the suspended portion of the chains, including the bolts, &c. - -	1,200,000
The weight of the roadway -	1,300,000
Making the total weight suspended	2,500,000 $=$
1116 tons.	

158. The weight with which this bridge may be loaded, exclusively of its own weight, at the rate of 70 lbs. per square foot,
will be about 700 tons. Total, therefore, 1116 + 700 = 1816
tons. With a deflection of $\frac{1}{10}$th, the tension at the extremities
is 1·313 times the entire weight, or 1816 × 1·313 = 2384 tons
for the greatest strain on the iron at the points of suspension.
The section of iron is there 496 square inches, which will bear
without injury 496 × 9 = 4464 tons, or 2080 tons more than the
utmost strain in dead weight that it can be exposed to.

159. A very bold design was also proposed by Captain S. Brown. The dimensions of Captain Brown's
bridge were to be, —

	Feet.
Span - - -	780
Height above high water	220
Breadth of platform -	30
Carriage-way - -	20
Two footpaths, each -	5

The estimated cost - - - 30,000*l.*

FINDHORN BRIDGE.

160. This is a bridge now in course of erection, and nearly completed, under the direction of Captain S. Brown, R.N., over the River Findhorn, in the North of Scotland. The span is about 300 feet. The breadth of the platform is 23 feet: and the bridge is calculated to bear 250 tons in load, besides its own weight.

The main chains will contain twelve lines of bars 9 feet 8¾ long, out and out; 4 inches broad by 1⅛ thick = 49½ square inches of iron in all, united by coupling plates 18¾ inches long, and bolt-pins 2½ diameter. The coupling plates are 5¾ broad in the middle, and 7¾ broad across the ends, where the bolt-holes are drilled out, so as to allow 2⅜ metal all round the holes. The ends of the main chains will be held by' keys rounded on one side, 2 feet long, 3 inches broad, and 4 inches deep. The entire weight of iron-work suspended will be about 76 tons.

PROFESSOR MILLINGTON'S DESIGN FOR GREAT MARLOW
BRIDGE.

161. The construction of this bridge was originally
undertaken by John Millington, Esq., late Professor
of Natural Philosophy at the Royal Institution, who
made out a design, in 1829, for a suspension bridge
at Marlow, to communicate between the High Street,
Marlow, and the Bisham Road, in lieu of the old
wooden bridge and flood bridge.

Professor Millington's design was as follows : —
Main suspension pyramids. Four cast-iron towers,
set upon stone pedestals, 22 feet high above the tops
of those pedestals. (See the annexed sketch.) The

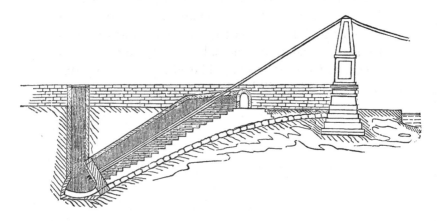

towers to be of the form shown in the sketch, and
of plates 2 inches thick, bolted together with 1¼-inch
bolts.

Main chains. The span or chord-line, from
centre to centre of the towers, to be 219 feet, *viz.*

the river is 184 feet wide; one of the pyramids was to
stand 10 feet away from the bank, and the other 15
feet, and their bases are 10 feet square.

The main chains to be eight in number, on each
side of the bridge, arranged in pairs, one under the
other, viz. two on each side of the bridge outside the
footpath, and two on each side of a central carriage-
way.

Each chain to consist of two lines of rods of hammered iron, 2 inches diameter, 10 feet long out and
out, connected by coupling links, and bolts $2\frac{3}{8}$ inches
diameter. The eyes in the ends of the chain-rods
to be bored out flat.

The total section of the chains is, therefore, $4 \times
\cdot7854 = 3\cdot1416 \times 16 = 50\cdot2656$ square inches. The
stretching strain is $452\cdot34$ tons; and the ultimate
strength 1357 tons.

The deflection of the main chains to be 16 feet.

The main chains to be carried over fixed cast-iron
saddles, laid in cavities in the upper parts of the pyramids, fixed to them by dovetailed tenons formed
on the ends of the saddles, fitting into dovetailed
mortises cast in the plates of the pyramids.

The back stays were to be carried down at an angle
of about 30 degrees, through a brick tunnel, about
55 feet long from the surface of the ground, built
on a broad foundation bed of stone-work, arched
gently from the fastenings, up to the pedestals at the
level of the ground.

The ends of the chains were to pass through a
cross wall of stone, about $3\frac{1}{2}$ feet thick, built across
the lower end of the tunnel, and were to be held by
bolts against a retaining-plate bedded against the
back of the cross wall.

L

162. The platform to be 30 feet wide, viz. one carriage-way 20 feet wide, and two footpaths 5 feet wide; to rise 2½ feet in the middle, from the horizontal line; to be 8 feet above the ground at the ends, and 12 feet above water-line, in the middle. To consist of cross joists, 12 inches by 5, suspended in pairs from the vertical suspending rods.

On these were to be laid eight longitudinal beams, 12 inches by 3, viz. one on each side of each footpath, and two bolted together on each side of the carriage-road. Over the cross joists, between the longitudinal beams were to be laid courses of 20 feet 3-inch deals. On these felt soaked with vegetable tar, and over that a course of 1½-inch planking for the footpaths, and of 2-inch planking for the carriage-road, spiked down to the under platform.

The parapet to be a wrought-iron railing, consisting of vertical standards connected by a top rail and diagonal bracing, and fastened down to a cast-iron plate bolted to the cross joists, extending all the length of the platform.

163. Professor Millington commenced the execution of this bridge, but relinquished it on undertaking the management of some mines in South America. The erection of the bridge was then placed in the hands of Mr. Tierney Clark, the engineer of the Hammersmith Bridge.

The bridge now erecting at Marlow, under his direction, is totally different from Professor Millington's design, and is nearly on the model of the Hammersmith Bridge, the main piers being archways of masonry. (See fig. 1., Plate VII., which is a front view of one of the main piers.)

The total length of the bridge, from end to end,

is 426 feet. The carriage-way is 20 feet broad; and the footpaths 5 feet broad each. The height of the platform, above the water, 12 feet. There are four main chains, composed of flat bars. The deflection is 18½ feet, and the section of iron in the chains 64 square inches.

NORFOLK BRIDGE, AT NEW SHOREHAM, BUILT AT THE
EXPENSE OF HIS GRACE THE DUKE OF NORFOLK.

164. This is a very beautiful bridge, now in course
of erection under the direction of Mr. Tierney Clark.

The general arrangement of the design is similar
to that of Marlow and Hammersmith Bridges. But
the architecture of the archways that form the main
piers is altogether richer and of greater beauty. (See
figs. 2. and 3., Plate VII., a front and end view of
one of the piers.)

The chord-line between the points of suspension is
284 feet, and the deflection of the chains 20 feet 2
inches = $\frac{1}{14}$th. The breadth of the platform within
the parapets, 28 feet 6 inches. The carriage-way 20
feet; and two footpaths 4 feet each.

There are three lines of chains on each side of
the bridge, containing in all a sectional area of 84
square inches. The bars are 8 feet $10\frac{1}{2}$ inches long,
centre and centre of the eyes, and $6\frac{1}{4}$ inches broad by
$1\frac{1}{8}$th inch thick in the body; at the eyes they are $8\frac{3}{4}$
broad. The coupling bolts $2\frac{1}{2}$ inches diameter. At
the points of suspension the chains lie on cast-iron
rollers, $10\frac{1}{2}$ inches diameter, with wrought-iron pivots
$2\frac{3}{4}$ diameter. The links which lie on the rollers, are
$6\frac{1}{2}$ broad in the body; and 9 inches at the eyes.

The platform differs rather from the ordinary construction. The cross bearers are of cast-iron, 14
inches deep in the middle, 10 inches deep where
the vertical suspending rods are fixed to them, and
thence diminishing to 6 inches deep at the ends;

they are 1 inch thick, with a flange at top and bottom, and they are not supported from the chains at their extreme ends, but only on each side of the central carriage-road; the footpaths, which are 4 feet wide, resting upon the projecting ends of the beams, beyond their points of support. The usual double set of suspension rods, with all the consequent parts, is rendered unnecessary by this arrangement; and Mr. Clark has found, by extensive experiments upon cast-iron beams, made for this purpose, that their strength can be sufficiently depended upon when thus applied to support the footpaths.

The top flanges of the cross bearers have knobs cast on them, for securing the roadway planking, which is of oak, 3 inches thick; and over that is laid another course of timber, alternately flat and endwise, of the grain upwards.

In this bridge (and the Marlow Bridge) there is a central arched opening through the main piers, for the carriage-road; and the footpaths are turned outside of the piers (as shown in fig. 2.), projecting out beyond their sides, and supported upon proper cast-iron consols.

The abutments are of solid work, consisting of brick and stone-work, weighing each 900 tons. The backstays go off from the main piers at a different angle to the central chains, passing through tunnels in the abutments, and are fastened, at the back of them, against large cast-iron plates, by oval wrought-iron bolts, 7 inches by 3 inches.

The length of the platform is 268 feet, and the total weight computed to be suspended in the central opening, including the supposed load at the rate of 62 lbs. per square foot of platform, is 356 tons.

SECTION VI.

REMARKS ON MATERIALS USED FOR SUSPENSION BRIDGES.

Comparison of Wire with Bar-iron as a Material for the Supporting Chains of Suspension Bridges.

165. SOME of the French engineers appear to prefer wire to bar-iron for suspension bridges. The reasons they give are, that iron-wire is stronger than bar-iron; that cables of wire can be put together more easily and rapidly than chains; that it is more easy to ascertain whether it be sound or not; and, lastly, that it is an easier operation to get wire cables up into their places than bar-chains.

166. The facility of working wire without heavy machinery, and almost without any but the most common tools, may render its use expedient under certain circumstances, viz. for a small bridge, and where the engineer is deprived of mechanical aid; but it has many disadvantages, which would become glaring if it were applied in the construction of a large and strong bridge.

First, although a single wire is stronger *per square inch* than a bar of iron, it is much to be doubted whether a *cable* made of wires is stronger than a bar-chain of equivalent dimensions, because of the inequality of tension in the several wires, which throws a greater share of strain on some than on others, and, therefore, reduces the effective strength of the cable to that of a cable of less diameter. This inequality it is hardly possible to prevent, even if the wires are drawn, in making up the cables, to the same curvature that they are intended to have when in their places.

Secondly, wires exposing a surface greater than bars of equal section, are more quickly destroyed by oxidation. A coating of varnish, it is true, may somewhat preserve them; but bars may also be preserved by varnish and painting, and are still, on that point, superior to wire cables.

Thirdly, wires are very apt to have kinks and bends in them, which cannot be got out without a very considerable strain*; and, when that has been done, it is difficult to ascertain whether the wire has not been permanently injured at that part. The long bends, also, that are formed frequently in wire, can hardly be got rid of at all. The author has repeatedly strained wire something less than $\frac{1}{10}$th inch diameter, (the ultimate strength of which would be about 600 lbs.) to a tension of $\frac{1}{17}$th of the chord-line, and then loaded it in the middle with a weight of 130 lbs., and caused that weight to be jerked upon the wire, so as to produce a strain upon it very little short of its breaking strain, without being able to remove the bends that had formed in it. Wire will, in fact, often break before losing its bends; and yet, if they are allowed to remain, be it ever so little, they of course impair the equality of strain upon the several wires that form the cable.

Lastly, although a small cable is very easily got up into its place, it is so, not because it is a cable,

* Mons. Vicât states, in speaking of No. 18. wire, the ultimate strength of which was 1165 lbs., that it required sometimes from 116 lbs. to 350 lbs. to take out the bends.

but because it is comparatively small and light. But a cable of large dimensions, say 3 inches diameter, or more, would be by no means more easily managed in raising it than a bar-chain of equal section. The latter would, on the contrary, be the more flexible of the two.

167. The ductility and manageableness of wire in working it into cables, and in handling the cables when made, have been also much exaggerated. Mons. Vicât, who, in building the bridge of Argentât, had some experience in the use of wires, allows this, notwithstanding his final preference of wire. He says, " I had accustomed myself so completely " to look upon the flexibility of wire cables as like " that of hempen cables (that is, for moderate bends,) " and all I had read on the subject gave me so great " a feeling of security, that I had never even given a " thought to the effect that would be produced on " the equal tension of the wire, by the new curves " that the cables would assume when raised into their " place. This effect is, however, immense."

Elsewhere he says, after stating the defects of wire: " It may be taken as a standard, that a cable, properly " bound round, must be considered as having the " rigidity of a bar of iron. Consequently, such a " cable cannot be rolled up, nor bent backwards and " forwards. And, moreover, if it is beyond a certain " length, and that not a very great one, it is hardly " practicable either to move it about, or to raise it up " into its place." *

168. In fact, if wire must be used, it would be better to form it into links of from 10 to 15 feet

* Déscription du Pont Suspendu à Argentât, par L. T. Vicât, &c., p. 18.

long, and couple them with short links, either of wire or iron, and bolts of large diameter, made hollow for lightness, in the way described, p. 117., for the *Geneva Bridge.*

All things, however, considered, it may be safely pronounced, that bar-chains are better adapted than wires for any thing beyond the size of a foot-bridge.

*Of the Use of Wood for the Supporting Chains of Suspension
Bridges.*

169. The materials of which a suspension bridge
is constructed, as well as its proportions, ought to
depend upon the purposes for which it is intended.
Iron is undoubtedly the best material for large
bridges, destined for constant and heavy traffic; and in
this country iron is so cheap, and so readily procured
in almost every part of the kingdom, that it will
rarely be desirable to seek any substitute for it.

But bridges are sometimes erected in remote
places, particularly abroad, where iron and mechanics
are not easily or cheaply procured, and where all that
is wanted is an established passage, but not a bridge
to bear any great load. It may, therefore, be useful
to consider how far, on such occasions, wood could be
substituted for iron.

170. Wood, used as a material for making the chains
of a suspension bridge, is in some respects superior,
in others inferior, to iron.

The advantages are, that, weight for weight, its
absolute strength is greater than that of iron, and
that it is worked more readily and with fewer tools.

N.B. In the last section will be found tables of the strengths,
both longitudinal and transversal, of various woods. From them
it will be seen that the ultimate strength of cohesion of fir is
12,000 lbs. per square inch = 5·35 tons, or about $\frac{1}{3}$th of the
strength of iron.

The weight of fir is about $\frac{1}{12}$th of the weight of iron: hence, weight for weight, the strength of cohesion of fir is 2·4 times that of iron.

The strength of ash is 2·8 times that of iron.

ditto of oak 1·33 ditto ditto.

171. The disadvantages of wood are, that it is more likely to decay by exposure, and that it stretches more; viz. it has been found by experiment, that fir will stretch about 20 times as much as iron, the sections being equal, and the strains being in proportion to the strengths. Hence, as the section of fir would require to be about five times that of iron to give equal strength, it follows that, with equal ultimate strength, *main-chains made of fir bars would stretch about four times as much as iron chains.*

172. Many plans might be devised of fastening bars of wood together, so as to form a continuous chain of detached bars. They might be connected by iron coupling plates and bolts, like the bars of iron chains, a metal bush being put into the bolt-hole, to prevent the iron bolt working against the wood.

Or wood bars might be connected by forming their ends into either cones or wedges, and enclosing them between cast-iron sockets, made in two pieces put together over the ends of the bars, and held by an iron band or hoop clasped round them.

173. The following sketch shows a very simple method of fastening, which would be convenient for a military bridge, or any such temporary construction.

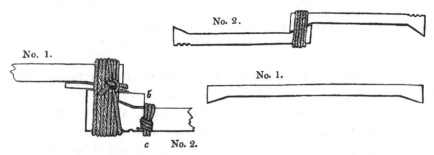

The bars of wood are to be made with tapering or wedge-shaped ends, and two of them being put together, as shown in the sketches, to be bound round with good strong rope. Iron straps, with keys and keepers, would form a more complete fastening, but of course more expensive, and could not be got ready with so little workmanship. The bars thus connected cannot be drawn away from each other without bursting open the fastening.

174. The author has tried this mode of fastening in a model 26¼ feet span. (See the sketch below.) The bars of wood were 14 inches long, out and out. Of two kinds, viz. No. 1. (see the sketch.) of good pine, ¾-inch square in the middle, with solid wedge ends 1⅛th inches deep, and 1½ long. Some of the bars No. 1. were bound together with seven turns of common cord, of barely ₁₆/₁₀ths of an inch diameter; others with iron wire: some had notches cut in the back, to receive the cord; others were left quite smooth.

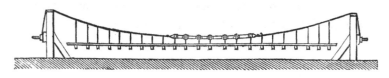

The bars (No. 2.) were of very common fir, 1 inch

deep by $\frac{3}{4}$ths in the middle, with solid wedge ends 2 inches deep and 2 inches long.

The bars No. 2. were all notched at the back, and bound together by 12 to 14 turns of cord.

Lastly, one of the fastenings was made with a bar No. 1. and a bar No. 2. put together, and as the inclinations of their wedge ends were different, a couple of loose wood wedges, *a*, *b*, were driven in between them. The ends of the bars were tied together with the $\frac{3}{16}$ths cord, a small piece of cord being also tied round at *c*, to prevent the loose wedge drawing out. The middle part only of the supporting chain was made of these wood-bars, as will be seen by the sketch of the model; viz. 4 or 5 feet in length on each side; the rest was made of iron wires.

The strains the model was tried with were as follows:—

TABLE OF EXPERIMENTS ON WOOD CHAINS.

Number of Experiments.	Span.	Deflection.	Load in Dead Weight.	Load falling on the Platform.	Momentum of the falling Weight.	Remarks.
	315 inches, or 26¼ feet.					
I.		26 inches.	none.	130 lbs. through 18 inches.	($\sqrt{1\cdot5 \times 8\cdot01 \times 130}$ lbs. =) 1274 lbs.	No injury.
II.		26 inches.	none.	130 lbs. through 33 inches.	1726 lbs.	One of the wood-bars torn in two.
III.		26 inches.	52 lbs.	130 lbs. through 34 inches.	1735 lbs.	No injury done.
IV.		22 inches.	52 lbs.	130 lbs. through 30 inches.	1645 lbs.	One of the wood-bars torn in two.
V.		21 inches.	52 lbs.	130 lbs. through 34 inches.	1735 lbs.	No injury.
VI.		26 inches.	4 persons, about 580 lbs.	none.	-	While the weight remained steady, nothing gave way. On swaying up and down, one of the wires broke. The fastenings remained sound.
VII.		26 inches.	Same, with additional 52 lbs. = 632 lbs.	-	-	While at rest, the bridge bore the weight. One person jumping broke one of the iron wires.
VIII.		28 inches.	none.	130 lbs. through 32 inches.	1697 lbs.	One of the wood-bars was torn in two pieces. The fastenings remained sound.

In the course of these trials, the fastenings of the bars, No. 1, drew considerably, but did not give way. Those of the bars No. 2., in which the inclination of the ends was more rapid, did not show any symptom of derangement.

SECTION VII.

ON THE THEORY OF SUSPENSION BRIDGES, AND ON
THE STRENGTHS OF MATERIALS USED IN CON-
STRUCTING THEM.

175. THE curve formed by a heavy flexible chain
hanging freely between two points is a catenary; but
when from the chains of a suspension bridge a heavy
platform is suspended at equal intervals, the curve of
the chains ceases to be a catenary, and assumes the
form of a parabola, or approaches it so nearly, that it
may be taken and computed as such.

Strictly speaking, the curve of the chains of a
suspension bridge is never either a catenary or a
parabola, but an intermediate curve between the two;
for it would be only a pure catenary if the weight of
the roadway were so inconsiderable in proportion to
that of the chains as to be taken for nothing; and if,
on the contrary, the weight of the chains were so
much less than that of the roadway that the former
might be taken for nothing, then only would the
curve be a pure parabola.

In practice, the curve is generally nearer to a
parabola than to a catenary, and may be treated as
such; all the computations depending on the nature
of the curve being much simpler when it is considered as a parabola, than when it is taken to be a
catenary.

176. The practical points involved in the nature of
the curve of the chains are : —

I. The *tension*, or strain, produced by the suspended weight on
any part of the length of the chains.

II. The exact deflection of the chains at any part from the horizontal chord line; because thence are ascertained the proper lengths for the vertical suspending rods.

III. The effect of the suspended bridge and its load, in producing horizontal draw and vertical pressure on the suspension piers.

The first and third depend on the angle of direction of the curve, that is, the angle that a tangent to it at the points of suspension forms with the horizontal chord-line.

Within the limits of those curvatures that are usually given in practice to the chains of suspension bridges, the angle of direction of a catenary and of a parabola are so nearly alike, that it is of little consequence under which head the curve is classed. Hence, tables of tensions, &c., calculated on the assumption that the curve is a catenary, may be used for a parabolic curve, in all that relates to *the effect of the weight in producing strain.*＊

The second point is one of greater nicety ; because the parabola and the catenary are so far different in curvature, that the deflections calculated for the one curve, will affect practically the lengths of the vertical suspending rods, if applied to the other. In reality, those lengths cannot be calculated with perfect accuracy, because, as before stated, the chains of a bridge are neither a pure catenary nor a pure parabola, and their curve varies in every case ; hence, when it is possible, the lengths of the vertical rods should be determined by experiment : when that cannot be done, they may be calculated by approximation, by a rule which will be given in another part of this section,

＊ The results will, of course, not be perfectly correct, but the errors will be trifling ; and a bridge ought never to be built without a superabundance of strength sufficient to compensate for them.

and the inaccuracies must be compensated for by practical means of adjustment.

Of the Tension of the Chains.

177. If a heavy chain were hung over two pillars of suspension, so as to form a loop, each side of which should be vertical, it is sufficiently obvious, that the tension, or strain on the iron at each point of suspension, would be just half the weight suspended.

But whenever the chain is drawn up nearer to a horizontal line, then the tension at each point of suspension is more than half the weight of the chain ; it is greatest when the chain is drawn nearest to a straight line, and least when it is most deflected therefrom.

Also the tension on the iron is different at every part of the length of the chain, whatever be its degree of curvature, being greatest at the extremities of the curve, or points of suspension, and least at its summit, or in the middle of the length of the chain.

Measure of Tension.

178. The tension of the chains at any point, whether the curve be a catenary or a parabola, is according to the angle of direction of that point ; *viz.*

The tension at any point is inversely as the angle that a tangent thereto makes with the horizontal chord-line.

The tension at each point of suspension is : —

The total length of the chain in feet or inches multiplied by the weight of 1 foot or 1 inch of the chain, divided by twice the sine of the angle of direction of the curve at the extremities ;

M

And is expressed by the following formula : —

$$A = \frac{h\,l}{2\,\sin.\,c}$$

A   representing the tension at the point of suspension,
h   the length of the curve in feet or inches,
l   the weight of a foot or an inch of the chain,
c   the angle between the tangent to the point of suspension
    and the horizontal chord-line *,
h l is equivalent to the whole weight suspended ; hence, when
    the formula is applied to calculate the tension at the
    points of suspension in a bridge, the angle c, and the
    whole weight suspended being known, it may stand thus:—

FORMULA I.

$$\left.\begin{array}{l}\text{The tension at the point}\\ \text{of suspension}\end{array}\right\} = \frac{\text{the whole weight suspended}}{2\,\sin.\,\text{angle of direction}}.$$

179. The results calculated by this formula correspond, as will be seen by the following table, with
those of experiments made by Mr. Rhodes, during
the erection of the Menai Bridge, to ascertain the
actual strain exerted in drawing up a chain weighing
3599 lbs., with a chord-line of 570 feet, to different
degrees of curvature. †

TABLE A.

Deflection in Parts of the Chord-line.	Angle of Direction of the Curve at the Points of Suspension.	Tension at the Point of Suspension found by Mr. Rhodes's Experiments.	Tension at the Point of Suspension computed by Mr. Barlow's Formula.
1 in 13·26	about 16°46′	1·7 times weight	1·73 times weight
1 in 11·7	do. 19°	1·51 do.	1·54 do.
1 in 14·2	do. 15°57′	1·82 do.	1·82 do.
1 in 15·4	do. 14°38′	1·95 do.	1·97 do.
1 in 16·28	do. 13°52′	2·04 do.	2·08 do.

* See in Barlow on the Strength of Timber, Appendix ; Investigation respecting Suspension Bridges.

† For a Table of Mr. Rhodes's experiments, see Provis's Account
of Menai Bridge, Appendix.

RULES FOR COMPUTING THE ANGLE OF DIRECTION OF THE
CHAINS OF A SUSPENSION BRIDGE, THE LENGTH OF
THE CHAINS, AND THE TENSION, CONSIDERING THE
CURVE AS A PARABOLA.

180. *To find the angle of direction of the curve
at the point of suspension, or the angle formed with
the horizontal chord-line by the tangent to that
point.*

Having given the chord-line and the deflection in
the middle, then,

FORMULA II.

The sine of the angle of
direction of the curve at
the point of suspension
$$\left.\right\} = \frac{2 \text{ deflection}}{\sqrt{(2 \text{ deflection}^2 + \text{semichord}^2)}} \cdot *$$

---

* DEMONSTRATION : —
Let $c\,b\,h\,l$ in the figure be the chain,
$c\,i$ the semichord,
$i\,b$ the deflection.

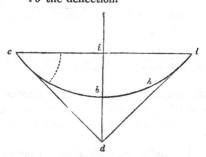

Double the deflection $i\,b$, making $i\,d = 2\,i\,b$; Join the points
c and d. Then $c\,d$ is the tangent to c, and the angle c is the
angle of direction of the curve at the point of suspension.
Now, because $c\,i\,d$ is a right-angled triangle, we have : —

Example. — Let the chord-line be 1000 feet,
deflection - 50 feet.

Then:

$$\text{sine } c = \frac{100}{\sqrt{500^2 + 100^2}} = \frac{100}{510} = 0\cdot196078 \text{ ; whence if the en-}$$

tire weight suspended be known :

$$\text{The tension at the point of suspension} = \frac{\text{weight}}{2 \sin . c} = \frac{\text{weight}}{\cdot392156}$$

$= 2\cdot55$ times the weight.

181. Generally : —

FORMULA III.

The sine of the angle of
direction of the curve
at *any point* x (see the
fig. following.)
$$= \frac{\text{twice the absciss of the point } x}{\sqrt{(2 \text{ absciss } x)^2 + \text{ordinate } x^2}} .$$

Hypothenuse $c\,d$ (or tangent c) $= \sqrt{c\,i^2 + d\,i^2}$; or since $c\,i$ is
the semichord, and $d\,i = 2\,i\,b$,

(1) Tangent $c = \sqrt{(2 \text{ deflec.}^2 + \text{semichord}^2)}$;

(2) Also ; sine c : its opposed side $(d\,i)$ :: sine i : its opposed side $c\,d$ (or tangent c).

And since $c\,d = \sqrt{(2 \text{ deflec.}^2 + \text{semichord}^2)}$, and the angle i
$= 90°$, the proportion (2) becomes:

sine c : 2 deflection :: 1 : $\sqrt{(2 \text{ deflec.}^2 + \text{semichord}^2)}$;

And,

$$\text{sine } c = \frac{2 \text{ deflec.}}{\sqrt{(2 \text{ deflec.}^2 + \text{semichord}^2)}} \text{ as stated in Formula II.}$$

* To find the absciss of any point x by calculation, when only
the chord line and deflection of the curve, and horizontal distance
of the point x from the point of suspension are known : —

The absciss of
any point x
on the curve
$$= \frac{(\text{ordinate } x)^2 \times \text{absciss of point of suspension}}{(\text{ordinate of point of suspension})^2} ;$$

because, in the parabola, the abscisses are proportional to the
squares of their ordinates.

Now the ordinate of any point x is:

And the tension at x = $\dfrac{\text{The weight suspended between the point } x \text{ and the middle of the chain}}{\text{sin. } x}$.

That is, in the figure:

$$\text{sine } x = \frac{2\,i\,b}{\sqrt{2\,i\,b^2 + x\,i^2}}$$

And the tension at x = $\dfrac{\text{weight of the length } x\,d\,b \text{ of the chain, and of whatever is suspended therefrom}}{\text{sin. } x}$.

182. This formula may be used for computing the tensions when they are tangential, *viz.* for any point where the direction of the curve forms an angle with the horizontal chord line ; but at the lowest point that angle becomes $= 0$, and the tension is there horizontal.

When the tension at the extremities or points of suspension is known :

The semichord $-\left\{\begin{array}{l}\text{the horizontal distance of the point } x \text{ from}\\ \text{the point of suspension ;}\end{array}\right.$

Or ordinate x, is the horizontal distance of the point x from the middle of the chord-line.

The ordinate of the point of suspension is the semichord, and the absciss of the point of suspension is the middle deflection. Therefore :

$$\text{Absciss } x = \frac{\left(\begin{array}{c}\text{horizontal distance of } x \text{ from}\\ \text{the middle of the chord-line}\end{array}\right)^2 \times \text{deflection}}{\text{semichord}^2}.$$

And these values of the ordinate and absciss of the point x being found in numbers, and used in the Formula III., will give the sine of the angle of direction of the curve at the point x, and consequently the tension thereon.

The tension at the $\left.\right\}$ lowest point $= \left\{\right.$ tension at extremity $\times$ cos. angle of direction at the extremity.

And as the tension at the extremity is $= \dfrac{\frac{1}{2}\ \text{weight}}{\sin.\,c}$, c being the angle of direction at the extremity, the tension at the lowest point may be expressed thus : —

FORMULA IV.

$$\text{Tension at the lowest point} = \frac{\frac{1}{2}\ \text{weight} \times \cos.\,c.}{\sin.\,c.}$$

183. The Formula III. for calculating the tension of the chains at any point is given, because it must be considered as belonging to the subject. But it would be very laborious, and very nearly useless, to compute the tension for any considerable number of points of the chains of a suspension bridge. In practice, it is not usual to make any regularly proportionate difference in strength between the parts of the chain ; the links are frequently made stronger where they pass over the saddles, and also at the fastenings in the abutments ; but the strength throughout that part of the chains which hangs between the suspension piers is generally uniform, and is proportioned to the greatest tension, *viz.* to the tension at the points of suspension.

The differences of tension at the different points on the length of the chains, are, in fact, so trifling, that it is not worth while attempting to save weight and metal by nicety of proportion, in a bridge of moderate dimensions. For example : —

TABLE B.*

With a Deflection of	The Tension at the middle or lowest Point is	The Tension at the Extremities is
1 in 39·97	4·992 times weight.	5·017
1 in 19·93	2·483	2·533
1 in 15·1	1·878	1·943

And the differences at any usual deflections are not such as to allow any great saving of materials in a large bridge.

184. Taking, for instance, the last of the Table for an example; suppose the Menai Bridge to have a deflection of 1 in 15·1, and the whole weight of the suspended part, with its fair load, to be 1000 tons. Then the tension at the middle would be 1878 tons, and at the extremities 1943 tons.

The chains of the Menai Bridge contain 260 square inches of iron; if, therefore, they were proportioned throughout to their tension, they should contain at the middle $\frac{260 \times 1878}{1943}$, or only about 251 square inches, instead of 260 square inches; difference 9 square inches. And as the strain increases continually from the middle upwards to the points of suspension, the strength of the chains must increase in the same

* These are taken from a very valuable Table for facilitating calculations relative to suspension bridges, computed by Davies Gilbert, Esq., V. P. R. S., read before the Royal Society, May, 1831, and printed in the Philosophical Transactions.

It contains the length of the chain in parts of the chord-line; the tension at the middle, and at the extremities, in parts of the whole weight suspended; and the angle of direction of the curve at the extremities; for 42 different deflections, from 1 in 39·97 of the chord-line, to 1 in 7, which are the extremes that are ever likely to be required.

ratio. Hence the whole saving on the length of the chains would not be above half the above difference, or, only 4½ square inches, which, multiplied by the length of the chains, 570 feet, is equivalent to a bar of iron, 2565 feet long and 1 inch square.

The weight saved would be, consequently, about 2565 × 3·3 lbs., (the weight of a bar 1 foot long and 1 inch square) = 3¾ tons, in a bridge which weighs 643¾ tons without any load.

It is obvious, therefore, that in a bridge of moderate size, so very perfect a proportion between strength and strain in every part would be a needless refinement in practice.

185. In a very large bridge, the weight that can be saved by proportioning the size of the bars to the tension at different parts of the length of the chains, is worth notice.

In Mr. Brunel's bridge at Clifton, for instance, it is proposed to make the chains, at the points of suspension, with 496 inches sectional area. Thence, they diminish towards the middle, where they will contain 458 square inches. The weight saved will be somewhere about 20 tons, which may be advantageously disposed of in other parts.

186. The tension of the chains at the extremities or points of suspension, may also be computed by the following simple formula, which does not require the use of tables of sines.

Formula V.

$$\text{The tension at the points of suspension} = \frac{\left(\sqrt{\text{semichord}^2 + 2\ \text{deflection}^2}\right) \times \text{whole weight suspended}}{4\ \text{deflection.}}$$

Example.—Let the semichord be 500 feet,
        the deflection   50 feet.

Then:

$$\text{Tension at points of suspension} = \frac{(\sqrt{500^2+100^2}) \times \text{weight}}{200}$$

$$= \frac{(\sqrt{26,000}) \times \text{weight}}{200} = \frac{510 \times \text{weight}}{200}, \text{ or } 2\cdot55 \text{ times the weight}$$

suspended.

187. The results obtained by the Formula V. accord with Mr. Rhodes's experiments (as will be seen by the following Table), with sufficient accuracy to justify its use in practice.

TABLE C.

Comparing the results obtained by Formula V. with Mr. Rhodes's experiments.

Chord-line.	Deflection.	Tension at Extremities, by Rule.	Tension at Extremities, by Experiments.
570 feet.	40 feet.	1·85 times weight.	1·82 times weight.
570 —	37 —	1·99	1·95
570 —	35 —	2·09	2·04
570 —	49 —	1·5	1·51

188. The reason of the Formula V. is the same as that of Formula II., *viz.* that the effect of the weight on the points of suspension, may be computed by supposing it all collected in one point *e*, (see the figure following,) the intersection of the tangents A *e*, B *e* to the angles of direction of the extremities.

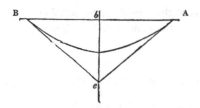

The strain on either of the points, A or B in the figure, is :

$$\frac{\text{Its tangent} \times \frac{1}{2} \text{ weight}}{b\,e}$$

The strain on the point A, for instance, $\Big\} = \dfrac{\text{A}\,e \times \frac{1}{2}\,\text{weight}}{1\,b\,e} = \dfrac{\text{A}\,e \times \text{weight}}{2\,b\,e}.$

Now $b\,e =$ twice the deflection of the curve ; and the tangent A e,

$= \sqrt{(2\,\text{deflect.}^2 + \text{semichord}^2)}$ (vide *antè*, equation (1), p. 164.)
    Whence :

$\dfrac{\text{A}\,e \times \text{weight}}{2\,b\,e}$ becomes $\dfrac{\sqrt{(2\,\text{deflect.}^2 + \text{semichord}^2)} \times \text{weight}}{4\,\text{deflection}}$, as

stated in Formula V.

It is on the foregoing property of the curve, that an approximative rule has been founded, which is sometimes used in practice, *viz.*

Formula VI.

The tension at the points of sus-
    pension   -   -   -$\Big\} = \dfrac{\text{chord-line} \times \text{weight}}{8\,\text{deflection}}.$

This rule is not correct; because the *tangent* of the angle of direction is the proper measure of the tension, and not the *semichord*.

When the deflection is not more than $\frac{1}{15}$th of the chord-line, the difference between the tangent and the semichord is not great; and hence the error of the Formula VI. is not very marked.

For instance, suppose the chord line = 300 feet,

deflection = 20 feet,

weight = 100 tons.

Then the semichord = 150 feet,

And the tangent to the $\left.\begin{matrix}\\ \\\end{matrix}\right\} = \sqrt{40^2+150^2} = 155$ feet.
angle of direction

Therefore Formula V. would give :

$$\text{Tension} = \frac{155 \times 100}{80} = 193\tfrac{3}{4} \text{ tons.}$$

And Formula VI. gives :

$$\text{Tension} = \frac{100 \times 300}{160} = 187\tfrac{1}{2} \text{ tons.}$$

The tension by Davies Gilbert's Tables for a deflection of 1 in 15 would be about 193 tons.

But with a great deflection the error of Formula VI. becomes more conspicuous.

Suppose the chord line = 350 feet,

deflection = 50 feet,

weight = 500 tons.

Then by Formula V. :

$$\text{Tension} = \frac{201\cdot5 \times 500}{200} = 503\tfrac{3}{4} \text{ tons.}$$

But by Formula VI. :

$$\text{Tension} = \frac{350 \times 500}{400} = 437\tfrac{1}{2} \text{ tons.}$$

The tension by D. Gilbert's Tables, with a deflection of 1 in 7, is 0·99 times the weight = (500 × ·99 =) 495 tons in this case.

189. *When the chord-line and deflection are given, but the length of the chain is not known ; then to find the length of the chain by approxi mation : —*

Formula VII.

$$\sqrt{\left(\text{Deflection} + \frac{\text{deflection}}{3}\right)^2 + \text{semichord}^2} = \left\{ \begin{array}{l} \text{Semi-length of} \\ \text{chain.*} \end{array} \right.$$

Example I.—Let the chord line be 1000 feet,
          the deflection   50 feet.

Then $\sqrt{(50 + 16\cdot66)^2 + 500^2} = \sqrt{4443 + 250,000} = 504\cdot4$ for the semi-length of the chain.

Mr. Barlow computes the length of the chain as a catenary, 1008 feet, for a chord-line of 1000 feet, with a deflection of 50 feet.

Example II. — Menai Bridge, chord-line 579 feet 10½ inches,
          deflection 43 feet.

Then $(43 + 14\cdot33 =) 57\cdot33^2$; and $\sqrt{289\cdot95^2 + 57\cdot33^2} = \sqrt{87317} = 295\cdot4$ for the semi-length of the chain, and the whole length $= 590\cdot8$ feet.   The length of the chains, by measurement from centre to centre of the saddles, is 590 feet.

Strain on the Main Piers.

190. The strain on the main piers of a suspension bridge is, from the arrangements usually adopted in practice, chiefly vertical, tending to crush them down, and only a partial horizontal strain is left, tending to pull them inwards.

The following is an exposition of the principle on

* The reason of this rule is, that if one half of a parabola, C *i f h*, were drawn out to a straight line, it would be = the right line C K; or the hypothenuse of a triangle C K L, one of whose sides C L = the semi-chord; and the other side

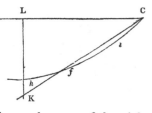

L K = ⅓ of the deflection L *h*; whence, because of the right angled triangle C K L, C K (or the half length of chain) = $\sqrt{(\frac{1}{3}\text{deflect.}^2 + \text{semichord}^2)}$, as stated in Formula VII.

which the suspended weight acts, to affect the stability of the main piers.

191. Suppose a light wire or a cord were passed over a pulley on the top of a column, as in fig. 1., and a weight of 100 tons hung at *a*; another weight of 100 tons must be hung also at *b*, to balance the weight *a*. And it is obvious, that the top of the column would be loaded vertically with 200 tons, tending to crush it downwards.

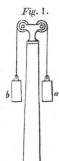

Fig. 1.

Now let the weights *a*, *b*, instead of being so disposed, be as in figure 2.

Fig. 2.

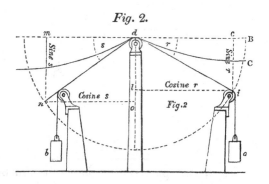

The effect of the weights *a*, *b*, will no longer be unmixed vertical pressure on the column, but they will each have a double tendency, *viz.* to pull the column over towards themselves horizontally, and to press upon it vertically.

The perpendicular pressure exerted by the weight *a*, for example, will bear that proportion to the whole weight *a*, that the perpendicular line *i c* bears to the radius *i d* of the arc B *i*. And the horizontal pull exerted by the weight *a*, will bear that proportion to the whole weight, that the horizontal line *i l* bears to the radius *i d*.

It is obvious, that the higher the point i rises through the arc towards B, (that is, the less the angle of direction B d i is made,) the greater is the proportion that i l will bear to the radius i d; consequently, the horizontal pull will increase, while the vertical pressure will decrease.

Finally, if the weights a, b, were disposed as in fig. 3., each end of the chains 2 being fastened to the summit of the column, they would exert each a purely horizontal pull upon it; and if the two weights were equal, they would negative each other, and the column would theoretically be unsolicited in any direction.*

Fig. 3.

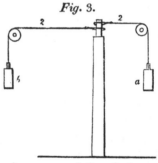

In practice, that part of the chains of a suspension bridge which is suspended in the central opening can never be horizontal, and the backstays are very rarely vertical. The strain on the main piers is, therefore, in general triple, *viz.* they are solicited inwards by the central suspended weight; outwards, by the tension of the backstays; and they are pressed vertically by both.

Measure of Horizontal Strain.

192. In applying the above principles to the heavy

---

* The wire or cord must be supposed to have itself no weight, since, if it had, no force could really strain it quite tight.

chains of a suspension bridge, loaded uniformly along
their length, and assuming the catenarian or parabolic
curve, we have to consider, not the effect of a weight
acting as shown in fig. 2., but the effort that the
piers would have to endure in horizontal pull, and
vertical pressure, from the *tension* of the heavy
curved chains.

That effort will be a proportion of the tension at
the points of suspension, variable according to the
angle of direction of the curve.

Fig. 2.

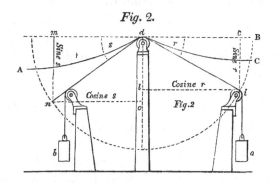

In fig. 2., let *d* C be half of the central chains
of a suspension bridge, and *d* A the backstays ; *r*
the angle of direction of the central chains, and *s* the
angle of direction of the backstays ; *i l* is the cosine
of the angle *r*, and *d i* the radius.

Now the horizontal pull upon the pier inwards will
(as stated in Art. 191. speaking of the effect of the
weight *a*) bear that proportion to the *whole tension*,
that the cosine *i l* bears to the radius *i d*.   That is ;

$$\text{Horizontal pull inwards} = \frac{\text{tension} \times \text{cosine } r}{\text{rad. } i\,d}.$$

Which, taking the radius as unity, becomes generally :

$$\left.\begin{array}{l}\text{Horizontal pull}\\ \text{inwards}\end{array}\right\} = \left\{\begin{array}{l}\text{tension} \times \text{cosine angle of direction of the}\\ \qquad\qquad\text{central chains.}\end{array}\right.$$

The horizontal pull outwards, is, in like manner, governed by the angle of direction of the backstays; *viz.*

The horizontal pull outwards of the backstay *d* A, fig. 2.
$$= \frac{\text{tension} \times n\,o}{\text{rad.}\,d\,n}.$$

Or, generally :

$$\text{Horizontal pull outwards} = \left\{ \text{tension} \times \text{cosine angle of direction of back stays.} \right.$$

193. It follows from this, that when the backstays are carried down from the summits of the suspension pillars vertically into the foundations (in the way described for the Geneva Bridge), the horizontal draw inwards, whatever it is, is wholly unbalanced. For then ;

$$\text{The horizontal draw inwards} = \text{tension} \times \text{cosine angle of direction of chains;}$$

And the horizontal draw outwards $= \text{tension} \times \text{cosine } 90° = 0.$

It is, therefore, the worst possible way of disposing the backstays, and should never be adopted if it can be avoided.

On the contrary, the best disposition is to carry the backstays away at the same angle, or very nearly so, as the central chains (as in the Menai and Hammersmith Bridges); for then the horizontal strains balance each other, and leave only vertical pressure on the piers, which they are the most capable of enduring.

Measure of Vertical Pressure.

194. The vertical pressure on the piers is measured by the *sines* of the angles of direction of the central chains and backstays ; *viz.*

The vertical pressure of the central part of the chains $\left. \right\} = \left\{ \right.$ tension × sine angle of direction of the chains ;

And the vertical pressure of the backstays $\left. \right\} = \left\{ \right.$ tension × sine angle of direction of the backstays.

Example. — Menai Bridge. — Call the whole suspended weight 1000 tons ;

The tension at the points of suspension for that deflection is, by Davies Gilbert's Tables, $\left. \right\} = (1000 \times 1\cdot749 =) 1749$ tons.

The angle of direction of the main chains is 16° 28′; the cosine of which is ·95898. Whence the horizontal pull of the central portion of the Menai Bridge, on the summit of each pier, is 1749 × ·95898 = 1677 tons.

The backstays are carried away very nearly at the same angle as the central chains ; whence the horizontal pull outwards on the main piers is equal to the horizontal pull inwards, leaving only, or very nearly only, vertical pressure upon them. But if they be examined, (see Plate IV.) it will be visible, by their form and dimensions, that they would bear a very considerable unbalanced horizontal pull in either direction, before giving way.

The vertical pressure of the central part of the bridge on the piers is 1749 × sine 16° 28′ = (1749 × ·28346 =) 495 tons, and the vertical pressure of the backstays the same; since, as before stated, the angle of direction is about the same. Total, therefore, 990 tons.*

---

* It is usual to carry the backstays down at an angle varying between 45° and the angle of direction of the central chains. It is not advisable much to exceed these limits; but if localities make it necessary to do so, the horizontal and vertical strains brought on the piers may be calculated by the foregoing formulæ, and their strength made proportionate.

195. The main piers should always be founded very solidly, and be of sufficient mass not to depend upon nice adjustment of the strains for stability.

When they are built in the river, or on soft ground, they should be erected on a platform laid upon piles, driven a good depth into the ground, and not more than $2\frac{1}{2}$ or 3 feet apart. If main piers are built in a river with a good rock foundation, the rock should be excavated to a slight depth, so as to lay part of the lower course of masonry a little below the surface.

A good plan for foundations, when the ground is loose and sandy, is to build upon wells, in the way practised in Madras for the public buildings.

These wells are made circular, about 3 feet diameter, and one brick thick. The first course is laid and cemented together on the surface of the ground; when it is dry, the earth is excavated inside and round about it, to allow it to sink. Then another is laid over it, and again sunk. The well is thus built downwards, sinking the brickwork bodily to a depth of 10 feet, or more, according to the building to be erected upon it, and the interior is filled up with rubble work. All the public buildings at Madras are erected upon foundations of this kind, which are found to answer very well.

196. The following table is computed by the formulæ in the preceding part of this section, and contains the proportions that are most required in calculations relating to suspension bridges, for those deflections that are of frequent occurrence in practice.

197. *Table exhibiting the Tensions of the Chains, Length of the Chains, and Sines and Cosines of the Angles of Direction, with given Deflections.*

TABLE D.

Deflection in the Middle, in Parts of the Chord-line.	Length of the Chain, in Parts of the Chord-line.	Tension at the Middle, in Parts of the whole Weight suspended.	Tension at each Point of Suspension, in Parts of the whole Weight suspended.	Sine of the Angle of Direction at the Extremities.	Cosine of the Angle of Direction.
$\frac{1}{15}$th	1·015	1·877	1·943	0·25733	0·9663
$\frac{1}{14}$th	1·018	1·753	1·823	0·27473	0·9615
$\frac{1}{13}$th	1·02	1·625	1·7	0·29415	0·95579
$\frac{1}{12}$th	1·0246	1·49	1·572	0·31805	0·94805
$\frac{1}{11}$th	1·0288	1·373	1·463	0·34176	0·93979
$\frac{1}{10}$th	1·0349	1·252	1·349	0·3714	0·92849

It must be remembered, that the lengths of the chain are computed on the assumption that their curve is a parabola, and will be only approximations nearer to or farther from the truth, according as the curve is more or less parabolic.

The foregoing table is constructed on the model of that cited in p. 162., by Davies Gilbert, Esq., and agrees with it very nearly in all the results,.except the lengths of the chains, which, by that table, are less than when computed by Formula VII. In the example taken in p. 166., the length of the chains by Davies Gilbert's tables, would be   -   -   588·297 feet.

By Formula VII. we found the length = 590·8 feet.

And it is by measurement   -   -   590 feet.

198. The degree of curvature that is given to the chains of a suspension bridge must, of course, depend much upon local and other circumstances. But when the engineer is free to choose his proportions, it is not advisable to give to a large and heavy bridge a deflection less than $\frac{1}{15}$th of the chord-line ; $\frac{1}{14}$th, or $\frac{1}{13}$th

is better. With these deflections, the chains are sufficiently taught to resist vibration and swinging, without being strained so as to lose too much of their effective strength. But, when these limits are much exceeded in either direction, the chains will be either overstrained, or too slack and liable to great vibration. Very light bridges, that have little to bear, may be made with a deflection as low as $\frac{1}{20}$th of the chord-line; but strength is lost by it, and nothing is gained, not even beauty, since a tolerably free and open curve is far more graceful and beautiful to the eye than a very flat one.

199. *To compute by approximation the deflection of the chains at any point, having given the chord-line and middle deflection :* —

Let x represent the difference between the middle deflection, and the deflection of any other given point a, whose horizontal distance from the middle of the chord-line is known.

Then : —

FORMULA VIII.

$$x = \frac{\left.\begin{array}{l}\text{middle de-}\\\text{flection}\end{array}\right\} \times \left(\begin{array}{l}\text{semichord}-\text{horizontal distance of the}\\\text{point } a \text{ from the middle of the chord-line}\end{array}\right)^2}{\text{semichord}^2} \text{ *}$$

---

* The reason of this rule is, that when we consider the curve of the chains as a parabola, the semichord is the *ordinate* to the point of suspension ; and the middle deflection is the *absciss* of that ordinate.

As the abscisses are proportional to the squares of their ordinates, the absciss of any other point y of the curve will be expressed thus : —

(No. 1.) absciss $y = \dfrac{(\text{ordinate } y)^2 \times \text{absciss to point of suspension}}{(\text{ordinate to point of suspension})^2}$

Now the *ordinate* to any point y is (as stated in the note to

Example. — Let the middle deflection be 50 feet,
    and the semichord - 500 feet.

Required the deflection of a point *a*, 400 feet from the point
of suspension, consequently 100 feet from the middle of the
chord-line.

Then $x = \dfrac{50 \times 100^2}{500^2} = \dfrac{500,000}{250,000} = 2$ feet for the difference between the middle deflection and the deflection of the point *a;*
whence the deflection of the point *a* is $(50 - 2 =)$ 48 feet.

At a point 200 feet from the middle of the chord-line, *x* would
be $= \dfrac{50 \times 200^2}{500^2} = \dfrac{2,000,000}{250,000} = 8$, and the deflection at that point
42 feet.

200. The practical use of Formula VIII. is to find
the proper lengths for the vertical suspending rods, by
computing the deflections of the chain at those points
where the vertical rods are to be suspended ; then
when the length of the *middle vertical rod* has been
settled, the lengths of the others will be found by
subtracting successively the deflections computed by
the formula from (the middle deflection + the length
of the middle vertical rod).

Suppose, for example, in the following figure, the
deflection *a b* = 43 feet ; the deflections *a c, a d,* &c.
to have been computed = 39 feet, 34 feet, &c. ; and
that it has been determined to make the middle rod
b g, 3 feet long.

Formula III.) the semichord — the horizontal distance of the
point *y* from the middle of the chord-line.

And the *absciss* of that ordinate is the middle deflection — the
deflection of the point *y;* whence the equation (No. 1.) will be
expressed in terms of the chord-line and deflection, thus : —

The middle
deflection
— deflec-
tionofany
point *y*
$\left.\begin{array}{l} \\ \\ \\ \\ \\ \end{array}\right\}$
$= \dfrac{\left(\begin{array}{l}\text{semichord — horizontal distance of the} \\ \text{point } y \text{ from the middle of the chord-line}\end{array}\right)^2 \times \text{middle deflect.,}}{(\text{semichord})^2}$

as stated in Formula VIII.

Then :

ch, will be $(43 + 3) - 39 = 7$ feet long.

di, $(43 + 3) - 34 = 12$ feet.

And so on.*

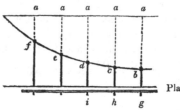

201. If the platform is to rise in any regular curve towards the middle, the increments of its rise must be first successively added to the sum of $(a\,b + b\,g,)$ and the deflections successively subtracted from the sum.

If, for example, the platform is to rise 3 inches for every successive vertical rod towards the middle, then $a\,b + b\,g$ being 46 feet in the above example,

ft.	in.	ft.	ft.

ch, will be $(46 + 3 =)\ 46{\cdot}25 - 39 = 7{\cdot}25$ long.

di, $(46 + 6 =)\ \ \ 46\tfrac{1}{2} - 34 = 12\tfrac{1}{4}$.

Of the Abutments, or retaining Piers.

202. It must be sufficiently obvious, that the stability of the whole bridge depends on the solidity of the abutments, since the ultimate pull is upon them. They should, therefore, always be as solid and as heavy a mass as possible, consistently with a rational regard to economy and appearance.   But of all parts

---

* The deflections and lengths of the vertical rods calculated by Formula VIII. will be only approximations; but the inaccuracies will not be greater than can be conveniently corrected by practical means of adjustment.

of a suspension bridge, the abutments are the last in which strength should be sacrificed to trifling economy. When the shores do not offer natural solid rock, and abutments are to be built of masonry, their mere weight should not be much less than equal to twice the utmost strain that can be brought on the chains by dead weight, and their total resistance (that is, in weight and resistance of adhesion combined) should be, at least, equal to 4 times the utmost strain.*

203. Mr. Telford, in his first design for Menai Bridge, proposed, as we have seen, p. 56. to attach the

---

* The standard assumed by the author for the *utmost dead weight* with which a bridge can be loaded, has been already stated, *viz.* 70 *lbs. per square foot of platform,* or one person for every two square feet; rating each person on an average at 140 lbs. weight. Some of the English engineers proportion their bridges at the rate of about 60 lbs. per square foot of platform.

The proof load that suspension bridges are subjected to by the government in France is about 41 lbs. per square inch ; and some of our suspension bridges in England, which have stood in constant service for more than ten years, are proportioned at the rate of not more than 26 lbs. per square foot ; that *is,* if they were loaded with more, the strain on the chains would be above the stretching strain of iron, or more than 9 tons per square inch. Our largest and strongest bridges, on the other hand, are proportioned above the standard of 70 lbs. per square foot.   If the Menai Bridge, for instance, were loaded at that rate, the entire strain on the main chains would be about 2000 tons ; while the chains containing 260 square inches of iron would bear, at 9 tons per square inch, 2340 tons, without stretching.

The question is chiefly one of expense, and resolves itself into the consideration, of whether a bridge shall be made so strong as not to be injured by a load that may come upon it, but is not likely so to do; or whether it shall be made only strong enough for its probable duty, taking the chance of failure, should it be tried by the utmost load it can hold.

main chains to the masonry of the land arches, in such a way, that to disturb the abutments they must have torn up a mass of masonry weighing about 12,000 tons, or six or seven times the utmost load that could come on the bridge.

The resistance offered by the adhesion of a mass of masonry to the ground that surrounds it, depends, of course, upon the nature of that ground, and cannot be very accurately determined by any general rule. For that reason, it ought not to be very much counted upon, unless the ground is manifestly so solid as to be almost rock ; and then it is better to make use of it as an abutment, carrying the chains through tunnels in the way described for the Menai Bridge and Brighton Pier, and fastening them to very large retaining stones, or cast-iron plates.

The foregoing observations are intended to apply only to the construction of large bridges. In small ones, good large stones, or cast-iron retaining plates, loaded with rubble work and earth, or held down by strong beams of wood, bolted to a few rows of piles driven close in front of them, will answer the purpose of abutments,

Strength of the main holding Bolts.

204. The retention of the chains in their places depends on these bolts. Therefore, their strength should be such, that they will not be injured permanently by the utmost strain that can be brought on the bridge. They must, in fact, be as strong as the main chains.

In the *Menai Bridge*, the holding bolts are twelve in number for each end of the main chains ; *viz.* (as stated in p. 57.) four bolts in the chamber of the middle tunnel for the two central chains, and four for each of the outer chains. Each of the sixteen chains is in effect held by a wrought-iron bolt, 6 inches in diameter, and $1\frac{1}{2}$ foot long between the bearings. (See the description, *antè*, p. 57.)

Now the strain that such a bolt would bear without flexure, if it were loaded *in the middle*, is

$$\frac{\text{(cube of diam.}=)216 \times 800}{1\cdot5 \text{ feet}} = \frac{172,800}{1\cdot5} = 115,200 \text{ lbs.} = 51\frac{1}{2} \text{ tons.}*$$

And as it is loaded pretty uniformly over its length, it will bear double as much, or 103 tons; and the 16 bolts would bear 1648 tons. They do not thus appear to be so strong as the chains, which would bear 2340 tons without stretching.

205. The abutments and the main holding bolts do not bear all the strain of a bridge, since the weight of the backstays balances a part of it. But, nevertheless, it is a good practice to make the retaining bolts of sufficient strength to bear the whole strain; because, as they are few in number, very little can be saved either in weight or expense by making them slight; and much risk may be incurred.

The failure of the Broughton Bridge was owing to deficiency of strength in the main holding bolts. (*Vide antè*, p. 93.) The deficiency is said to have arisen in part from bad quality in the iron; but that, if it were really the case, shows only an additional chance of failure; and an additional reason for giving a superabundance of strength to a part which, while it is more liable to undergo changes from being buried in the ground, or in damp tunnels, cannot be so easily and frequently examined as the other iron work;

* This calculation is founded on the assumption that a bar of good wrought iron, 1 foot long between the bearings, and 1 inch square, will bear, when loaded in the middle, 800 lbs. without flexure. See *post*, rules for strength of wrought-iron bolts, p. 195 to 199.

and which, lastly, by any failure, must inevitably derange, and, perhaps, destroy the whole bridge.

To calculate the proper Transverse Section for the Main Chains.

206. The section of the main chains of a suspension bridge must be proportioned so as to bear, without stretching, their own weight, the weight of the road-way, and the weight of the greatest load that can be brought on the bridge. In order to determine the sectional area of the main chains, *the chord-line, deflection, and length of the chains*, and also the *total weight to be suspended, exclusive of the weight of the chains themselves*, must be computed. These points being determined, then, *to calculate the proper section of the chains.*

RULE.

207. I. Multiply the length of the chain by 0·00148 : multiply the product by the number expressing the *tension* at the points of suspension, and subtract this last product from 9 for a *divisor*.*

---

* 3·3 lbs. or 0·00148 ton is the weight of a bar of wrought iron 1 foot long and 1 inch square, which multiplied by the tension and by the length of the curve gives the strain the chains endure per square inch from their own weight. And as a bar of iron 1 inch square will bear 9 tons without stretching, it follows that it will bear, *exclusively of its own weight*, the difference between 9 tons and the strain due to its own weight. The *divisor* in No. I. expresses therefore what the chains may be strained with *per square inch beyond their own weight ;* and if the strain produced by that additional weight (in roadway, vertical rods, people, &c.) be divided by the *divisor*, the quotient will be the number of square inches of iron that will resist the entire strain of load, chains and all, without stretching.

II. Multiply the whole weight that is to be suspended, exclusive of the weight of the chains, by the
tension; and divide the product by the *divisor* found.

The quotient is the section of iron that would exactly bear the weight of the bridge and load, without
stretching; that is, putting the rule into the shape
of a formula, it will stand thus : —

FORMULA IX.

$$\frac{\left(\begin{array}{c}\text{Weight(exclusive of}\\\text{weight of chains)}\end{array}\right) \times \text{tension due to the deflection}}{9 - (\text{length of chains} \times 0{\cdot}00148 \times \text{tension})} = \text{section of}$$

iron that will *exactly* bear the bridge and whole load without
stretching.

208. The chains must be made stronger than
that, more or less, according to the position of the
bridge.

For a bridge very much exposed to shocks and
vibration, the chains ought to be capable of bearing
from $1\frac{1}{4}$ to $1\frac{1}{2}$ times the utmost load that they are exposed to in dead weight; *viz.* they should contain a
sectional area $1\frac{1}{4}$ to $1\frac{1}{2}$ times that given by the rule.

If a bridge is not exposed to any violent shocks or
gales of wind, then $1\frac{1}{4}$ times the section given by the
rule, or even less, will be sufficient for the chains.

Example. — MENAI BRIDGE.

		Feet.	Inches.
209. Chord-line - - -		579	$10\frac{1}{2}$
Deflection - - -		43	0
Length of the curved chains		590	0

The tension at the points of suspension is 1·749 times the
weight suspended.

The weight of the roadway and iron work, exclusive of the

main chains, is about 398 tons; and the utmost load that can come on the bridge is 490 tons.* Total 888 tons.

Then to find the section of iron that will exactly resist the strain produced by a load of 888 tons, and by the weight of the chains themselves.

$$\frac{888 \times 1\cdot749}{9 - (590 \times \cdot00148 \times 1\cdot749)} = \frac{1553}{7\cdot473} = 207\cdot8 \text{ square inches.}$$

Now the weight of 207·8 square inches of chain 590 feet long, would be (207·8 × ·00148 × 590 =) 181½ tons, therefore the total suspended weight would be 888 + 181½ = 1069½ tons, which × 1·749 gives the strain = 1870½ tons. The stretching strain of 207·8 square inches = (207·8 × 9 =) 1870·2 tons.

Whence the chains of the Menai Bridge, if they contained only 207·8 square inches, would just bear without stretching, the utmost load that could come on the bridge. They really contain 260 square inches, or 1¼ times the section given by the rule; and would bear, without stretching, a strain of 347 tons more than the strain produced by the utmost load the bridge is capable of holding; viz. the weight of the bridge is 644 tons; the utmost load 490; total 1134 tons; producing a strain of (1134 × 1·749 =) 1993 tons. While the chains would bear, as before stated, a strain of 2340 tons without injury.

210. Formula IX. is not intended, as will be perceived by its construction, to decide what is the proper strength for a bridge, but only as a convenient mode of calculating *the section that will resist a given load.* What that load shall be, is the province of the engineer to determine, according to the position of his bridge, and the work it is appointed for.

There is, in fact, considerable difference in opinion and practice among engineers, as to the proper strength to be given to suspension bridges. Several

---

* The chord-line is 580 feet very nearly; but the suspended part of the platform is not above 560 feet long. Hence it contains (560 length × 28 breadth =) 15,680 square feet × 70 = 1,097,600 lbs. = 490 tons.

of the bridges described in this work do not possess
half the strength that would be assigned as proper
for them by the foregoing rule ; and others, again,
exceed it, while they have all stood unimpaired
hitherto, and have sufficed for their appointed work.

211. 70 *lbs. per square foot of platform* are assumed
as a standard for the load that may come on a bridge,
as being the utmost load that the platform could
hold, supposing it in fact quite filled with people,
crowded as close together as they could be.   This, it
is true, is not often likely to happen ; but it may do
so on a public occasion : and one overloading will
do injury as certainly, if not as extensively, as twenty.
Moreover, a bridge, without being half filled, may, if
the mass upon it be put in motion, be strained as severely as if it were quite filled with people at rest.

212. It is known that a body of men, whose
mere weight would scarcely be felt by a suspension
bridge, affect it severely by marching over it in
regular step. For, 1st. They must exert at every step
an extra force, or pressure, beyond their actual weight
upon the platform, to put themselves in motion ; and
that force, when they move quickly, is of the nature
of a sudden shock at every step. 2dly. They cause the
chains to vibrate, and then the weight of the chains
themselves thus put in motion adds to the strain upon
them ; and if the step of the men happen to chime
in with the undulations of the chains, the strain is increased to an extent very far beyond any thing that
could be produced by the mere weight.   The author
is not aware of the existence of any sufficient data,
drawn from experiments on a large scale, to form
correct calculations of the effect of the march of
troops on the platform of a suspension bridge ; but
he thinks it may be safely assumed,

213. 1st, That any body of men marching in step, say at 3 to 3½ miles per hour, will strain a bridge at least as much as double their weight at rest.

And 2d, That the strain they produce increases much faster than their speed, but in what precise ratio is not determined.

In prudence, not more than ⅙th of the number of infantry that would fill a bridge, should be permitted to march over it in step ; and if they do march in step, it should be at a slow pace.   The march of cavalry, or of cattle, is not so dangerous ; 1st, because they take more room in proportion to their weight ; and 2dly, because their step is not simultaneous.

Methods of raising the Main Chains.

214. It is difficult to give any precise directions for getting up the main chains into their places.

For a large bridge, a good way is to work upwards from the fastenings to the summits of the piers, putting the chains together on a scaffolding erected at the same inclination that the backstays are meant to have.*   Then to continue the chains over the saddles down in front of each suspension pier : that is, on one side, just to hang over the summit of the pier ; and on the other, to continue the chain down in front of the pier as low as the water, and then to continue putting it together on floats, or a raft, carrying it thus as far across the opening as its length will allow. Then to lay hold of the end of the chain on the raft by the *second link*, with tackle

---

* See account of raising the chains in the erection of the Menai Bridge, *antè*, p. 61.

carried over the summit of the opposite pier, and
worked by capstans on the shore, and so hoist up
the end of the chain of the raft till it reaches the
end of the chain hanging over the summit of the
opposite pier, in front of which a scaffold must, of
course, be erected for the people to stand on to put
the two end links together. The object of raising
the chain by the *second link,* is to have the end one
free and unstrained, that the workmen may raise it
easily and put the coupling bolts through it.

215. In adopting this mode of proceeding, it must
be remembered that the strain required to raise the
chains will equal, and perhaps exceed, the tension of
the chains themselves when placed ; and the hoisting
machinery ought to be strong enough to bear that
strain. For instance, if a chain is to be raised that
weighs 20 tons, and the deflection is to be $\frac{1}{15\cdot4}$th of
the chord-line, the strain required to raise it up
would be, according to Mr. Rhode's experiments
(*vide antè*, Table C, p.169.), 20 tons×1·95 tension
= 39 tons strain. To allow, therefore, for stiffness
of joints and other unavoidable sources of resistance,
every part of the hoisting machinery, capstans, ropes,
blocks, &c. ought to be proportioned to bear a strain
of half as much more, or about 60 tons.

216. The chains of many large bridges have been
put in their places on a temporary bridge, either of
ropes or of small chains ; and this is, perhaps, a
better way, where the position of the bridge is not
too much exposed, as there is less risk of injury to
the works from the failure of machinery, and the
operations are more divided and less heavy.

It would be next to impossible, and very nearly
useless, to attempt giving more than the outline of

the proceedings to be adopted in raising the main chains. * The details will vary in every case, and must be suggested, as difficulties present themselves, by the invention and presence of mind of the engineer. If he has not these qualities of mind, no instruction will avail him ; and if he has them, he wants no instruction that he could derive from a work of this kind.

* Those who wish to read a detailed account of the raising up of the main chains at Menai Bridge, may consult Mr. Provis's work, already cited, *antè*, p. 62., and will find in it a very complete one.

RULES FOR THE STRENGTH OF MATERIALS USED IN THE
CONSTRUCTION OF SUSPENSION BRIDGES.

Strength of Cast-Iron Beams.

217. The following is Mr. Tredgold's rule for the
strength of cast-iron beams.

FORMULA I.

$$\left.\begin{array}{l}\text{Weight that an}\\ \text{uniform cast-}\\ \text{iron beam will}\\ \text{bear in pounds}\\ \text{in the middle}\end{array}\right\} = \frac{\text{breadth in inches} \times 850 \text{ times square of}}{\text{length in feet between the supports}}.$$

Whence if the weight be known,

$$\text{The breadth in inches} = \frac{\text{length in feet} \times \text{weight in lbs.}}{850 \times \text{square of depth in inc.}}. \quad \text{And}$$

$$\text{The depth in inches} = \sqrt{\frac{\text{length in feet} \times \text{weight in lbs.}}{\text{breadth} \times 850}}.$$

When the weight is known, but neither depth nor breadth
are fixed, then Mr. Tredgold calls the breadth $= \dfrac{\text{depth}}{1}, \dfrac{\text{depth}}{2}$,
or any proportion n of the depth, and

$$\frac{(\text{the length in feet} \times n) \times \text{weight in lbs.}}{850} = \text{the cube of the depth,}$$

and the breadth is $\dfrac{\text{the depth found}}{n}$.

218. Mr. Farey, in his excellent " Treatise on the
Steam Engine," p. 613., gives a rule for the strength

o

of cast-iron beams deeper in the middle than at the ends : —

To find the proper dimensions for the middle part of a cast-iron beam, to sustain a given weight in the middle, when supported at each end, the depth of the beam at each end being *one third* of the depth in the middle.

If the breadth is fixed, then to find the depth in the middle,

Formula II.

$$\text{The depth in inches} = \sqrt{\frac{(\text{weight in lbs.}) \times \text{length in ft. between supports}}{250 \times \text{breadth in inches}}}.$$

If the depth be fixed : —

$$\text{The breadth} = \frac{(\text{weight in lbs.}) \times \text{length in ft. between supports}}{250 \times \text{square of depth in inches}}.$$

Example. — Let the load on the middle of a beam, the depth of which in the middle is to be 3 times the depth at the ends, be 56,000 lbs. ; the length of the beam between the supports 20 feet, and the depth in the middle 36 inches.

$$\text{Then the breadth} = \sqrt{\frac{56,000 \times 20}{250 \times 1296}} = \frac{4480}{1296} = 3.45 \text{ inches for breadth.}$$

If the breadth had been fixed = 4 inches,

$$\text{Then the depth} = \sqrt{\frac{56,000 \times 20}{250 \times 4}} = \sqrt{1120} = 33\frac{1}{2} \text{ inches for depth.}$$

That is, a beam 20 feet long between the supports, and four inches thick throughout, must be $33\frac{1}{2}$ inches deep at the middle, and 11·16 inches deep at each end, to bear constantly, without any risk, 56,000 lbs. in the middle.*

---

* In both the foregoing rules the load is supposed to be on the middle. If it be laid uniformly over the whole length, the beam will bear double as much as when it is accumulated on the middle. A beam, for example, of the above dimensions, loaded uniformly, would bear safely 112,000 lbs.

Mr. Farey's rule proceeds on the result of experiments made by Mr. Banks and by Mr. G. Rennie, showing that the *ultimate transverse strength* of a bar of cast iron, 1 inch square and 1 foot long between the supports, loaded in the middle, is 2600 lbs. The divisor 250, is therefore adapted to make cast-iron beams between 10 and 11 times stronger than the load they are intended for, which is the practice in making steam-engine beams ; but so great a measure of strength is not necessary in the parts of suspension bridges.

For instance, if the cross bearers of the platform are made of cast iron, they will be sufficiently strong if made capable of bearing from three to five times the load they are exposed to. The proper strength may be computed by Formula II., using from 500 to 850 for a divisor instead of 250, according to the degree of exposure of the beams to rough work.

Transverse Strength of Wrought-Iron Bolts.

219. In constructing steam engines and similar heavy machines, it is the practice of engineers to proportion all the moving parts, such as rods, links, cross heads, pins, &c. so that they shall not have to bear more than from $\frac{1}{12}$th to $\frac{1}{10}$th of the strain that would break or cripple them. For parts that are at rest, the strength is made about 3 or 4 times the strain.

The bolts that are used to connect the links in the chains of a suspension bridge are in a medium state between rest and motion. They are at rest when the bridge is merely loaded with a stationary weight, but whenever the bridge is caused to vibrate, they are put in motion and may be strained by jerks. Hence they ought to possess a greater strength than if they were perfectly at rest, and they are usually proportioned in practice as if they were moving parts.

Mr. Barlow, in his Essay on the Strength of Timber, says, that in experiments made by General Millar, at Woolwich, a bar of Swedish iron 1 inch square and 3 feet between the bearings, bore a load of 560 lbs. in the middle, and was deflected by it $\frac{1}{4}$ inch.

> With 716 lbs. the *deflection* was ·375 inch.
> With 884   -   -   -   ·5   inch.

Being then relieved of its load, the bar resumed a rectilinear form.

With 1120 lbs. the deflection was 1 inch, and the elasticity of the bar was destroyed.

As a mean, Mr. Barlow states 1000 *lbs. for the load that will destroy the elasticity of a bar of wrought iron, supported at both ends, 1 inch square and 3 feet long between the supports.*

220. Mr. Barlow does not state at what load flexure was first sensible; but if a bar was deflected $\frac{1}{4}$th of an inch by 560 lbs., it may be assumed that it would not be safe to load it permanently with more than 280 lbs. Taking, therefore, 280 lbs. as the strain that a bar of iron 3 feet long and 1 inch square will bear without injury, a bar 1 foot long and 1 inch square will bear 840 lbs. in the middle; or say 800 lbs. as a more convenient number for calculation, and on the safe side, since it will give a somewhat greater strength. On these data the following rule proceeds.

To find the proper diameter for a round wrought-iron bolt at rest, supported at both ends, and loaded in the middle.

RULE.—*Multiply the strain in pounds by the length of the bolts in feet between their supports; and divide the product by* 800. *The cube root of the quotient is the proper diameter for the bolt.** *If the bolt is to be loaded uniformly over the length instead of in the middle, then divide by* 1600 *instead of* 800.

Example.— In the *Menai Bridge*, each of the bolts that retain the 16 chains must bear 125 tons, since the whole strain on the chains may amount to about 2000 tons.† The length of each bolt between its bearings against the rock is 18 inches, and the chains are disposed along that length; therefore the divisor will be 1600.

Then:

$$\frac{(\text{strain in lbs.}=) \; 280,000 \, \text{lbs.} \times 1\cdot5 \; \text{feet}}{\text{of } 1600} = 262\cdot5.$$ The cube root of which is $6\frac{4}{10}$ inches for the diameter of the bolt; the actual diameter is 6 inches.

221. The foregoing rule is adapted to proportion the strength of a bolt to bear $3\frac{3}{4}$ times the strain it is exposed to before losing its elasticity, and will give sufficient strength for the *retaining bolts* of suspension bridges, which are at rest.

It is requisite and usual to make the *coupling bolts* of the chains stronger, for the reason stated in Art. 219. Their strength is usually in practice about

* Because the strength of bars of iron is inversely as the length, and directly as the breadths and squares of the depths, and therefore, when the bars are round, the strengths are as the cubes of the diameters.

† See computation of the weight that may come on the Menai Bridge,   Art. 209.

10 times the strain they may be exposed to from dead weight, and will be obtained by using the foregoing rule, with 300 for a divisor, if the bolts are to be loaded in the middle, and 600 when loaded as they really are, uniformly over their length ; the distance apart of the coupling plates *in feet*, being taken for the *length of the bolts between the bearings*.

Example I.

222. In the *Menai Bridge* every bar was proved to 11 tons per square inch, or 35¾ tons, and is adapted to bear that strain. The bolts must therefore be capable of bearing the same. The coupling plates are 1 inch apart.

Whence :

$$\frac{(35 \cdot 75 \text{ tons} =) \ 80{,}080 \text{ lbs.} \times \frac{1}{12}\text{th foot}}{600} = \frac{6673}{600} = 11 \cdot 1, \text{the cube}$$

root of which is 2¼ inches. They are really 3 inches diameter, and are stronger in proportion than other parts of the same bridge.

Example II. — *Hammersmith Bridge.*

223. The bars of the chains contain each 5 square inches of iron, and are adapted to bear 45 tons without stretching. The length of bolts between the coupling plates is 1 inch or $\frac{1}{12}$th of a foot.

Calculation : —

$$\frac{(45 \text{ tons} =) \ 100{,}800 \text{ lbs.} \times \frac{1}{12}\text{th foot}}{600} = \frac{8400}{600} = 14, \text{ the cube root}$$

of which is $2\frac{7}{16}$ inches for the diameter of the bolts; they are really $2\frac{10}{18}$ inches diameter.

Example III. — *Tweed Bridge.*

The bars of the chains contain each 3·14 square inches, and would bear, without stretching, 63024 lbs. The length of the bolts between the coupling plates is 2 inches.

Calculation : —

$$\frac{63024 \text{ lbs.} \times \cdot166 \text{ feet}}{600} = 17\cdot3, \text{ the cube root of which is } 2\cdot6$$

inches for the diameter of the bolts. Their real diameter is 2·16 inches. *

Strength of Cast-Iron Bolts.

224. We have seen (*antè*, p. 195.) that a cast-iron bar, 1 inch square and 1 foot long between the supports, will break when loaded in the middle with 2600 lbs.

If at rest, a cast-iron bar may be loaded with about ⅓th of that; and if it be a moving part, with about 1⁄12th.

Therefore, to find the proper diameter for a *cast-iron* circular bolt at rest, supported at both ends, and loaded in the middle :

FORMULA III.

$$\frac{\text{Load in lbs.} \times \text{length in feet between the supports}}{500} = \text{cube of diameter.}$$

If the bolts are to connect moving parts, then :

* From these examples it will appear that the foregoing rule is adapted to make the sizes of the bolts a little less than they have been made in the strongest bridges, and a little larger than they have been made in some of the specimens of slight construction. The author has little doubt but that it will give a strength for the bolts, proportionate to that of the other parts of a suspension bridge; but it must, of course, be taken and used with discretion, *viz.* if a bridge is much exposed, a less divisor than 600 should be used; if it is to be little exposed to shocks or vibrations, even a greater divisor may, perhaps, be used with safety.

Formula IV.

$$\frac{\text{Load in lbs.} \times \text{length in feet}}{220} = \text{cube of diameter.}$$

If the bolts are loaded uniformly over their length instead of in the middle, 1000 and 440 must be used as divisors, instead of 500 and 220.

Strength of Cast and Wrought-Iron Gudgeons.

225. The pivots, or gudgeons, of the great levers of steam engines are usually of cast iron, and are loaded at the rate of about 643 lbs. per circular inch of their area, the length of bearing of the pivots being 1¼ times the diameter (*vide* Farey on the Steam Engine, p. 605.), which is equivalent to 804 lbs. per circular inch, if the length of bearings were equal to the diameter. This standard is justified by long practice, the pivots of the great levers being scarcely ever known to break.

The cast-iron gudgeons of water wheels are usually loaded at the rate of about 500 lbs. per circular inch of their area.

Whence, as a general rule, to find the proper diameter for *cast-iron gudgeons* to bear a given weight in lbs., Mr. Farey gives the following.

If the length of bearing is to be equal to the diameter,

Formula V.

$$\sqrt{\frac{\text{weight in lbs.}}{500}} = \text{the proper diameter.}$$

If the length of bearing is not the same as the diameter, the weight must be multiplied by that length expressed in parts of the diameter.

For *wrought-iron gudgeons*, Mr. Farey gives the same formula, taking 1000 for a divisor instead of 500.

These rules give a very much greater strength than would be requisite for gudgeons not exposed to any shocks or sudden twists, because they are drawn from the practice in making the gudgeons of moving machinery, such as great levers of steam engines, water wheels, &c. ; and are consequently adapted to make the pivots like the other moving parts of heavy machinery, to bear at least ten or twelve times as great a load as they are exposed to ; whereas the pivots of rollers for supporting the main chains of a suspension bridge are at rest, and have only to bear absolute weight or pressure.

226. In applying Mr. Farey's rules to the pivots of supporting rollers for the chains of a suspension bridge, the load on the pivots will be the vertical pressure due to the tension of the central chains and backstays, and must be calculated by the formulæ in art. 192. Then, in computing the diameter of the pivots, Mr. Farey's rules may be used, taking from 1500 to 2000 for a *divisor* for *cast-iron* pivots instead of 500, and from 3000 to 4000 for a *divisor* for wrought-iron pivots instead of 1000.

Example.—Suppose, in a bridge, there are eight chains, each resting on four rollers with wrought-iron pivots. The length of the pivots to be equal to their diameter.

Suppose the utmost tension at the points of suspension to be 1600 tons; the deflection being 1 in 13, and the backstays carried away at the same angle as the chains of the middle opening.

Then, by Table D. (p. 179.), we have the vertical pressure on the summit of each pier, from the middle part of the bridge, $1600 \times \cdot 29415 = 470$ tons; and the same from the effect of the backstays; total, therefore, 940 tons.

The load on each pivot is therefore $\dfrac{940}{8 \times 4 \times 2} = 14\cdot7$ tons $= 32{,}928$ lbs.

Whence:

$$\sqrt{\frac{32{,}928}{3000}} = 3\cdot3 \text{ inches for the diameter of the pivots.*}$$

Strength of Wood.

Longitudinal strength.

227. The following is a table of the ultimate strength of cohesion of different woods, taken from Mr. Barlow's Essay on the Strength of Timber.

Name of Wood.	Load in Lbs. that will tear asunder 1 Square Inch.
Oak	10,000
Box	20,000
Ash	17,000
Teak	15,000
Fir	12,000
Pear	9,800
Mahogany	8,000
Beech	11,500

* This rule is not offered with great confidence in its correctness, as there are hardly a sufficient number of practical

It follows, that the ultimate strength of cohesion of a bar of any wood is its sectional area multiplied by the number in the table that expresses the cohesive strength of 1 square inch. Thus, a bar of fir 10 inches by 3 will be torn asunder with 12,000×10×3 = 360,000 lbs.

Mr. Barlow says, that wood ought not to be loaded with a strain in the direction of its length of more than two-thirds of its ultimate strength. One half would be still more advisable in practice.

Transverse Strength of Wood.

228. Mr. Farey, in his Treatise on the Steam Engine, gives the following rule for calculating the strength of oak beams : —

To find the proper dimensions for an oak beam to sustain a given weight in the middle, when it is supported at both ends.
When the load and the length between the supports are given, then if the *breadth* be fixed, to find the *depth*.

FORMULA I.

$$\text{The depth in inches} = \sqrt{\frac{\text{weight in lbs.} \times \text{length in ft. between supports.}}{40 \times \text{breadth in inches}}}$$

If the *depth* be fixed, then to find the *breadth*.

examples of supporting rollers in suspension bridges to ground it on the authority of established practice.

FORMULA II.

$$\text{The breadth in inches} = \frac{\text{weight in lbs.} \times \text{length in feet}}{40 \times \text{square of depth in inches}}.$$

229. Mr. Farey's rule may be used for finding the proper dimensions of beams made of any of the woods named in the following Table, by multiplying the divisor in Mr. Farey's formula by the number in the column of multiplicands corresponding to that wood.

Name of Wood.	Multiplicands to be used with Farey's Rule.*
English oak	1
Canadian oak	1·24
Ash	1·42
Elm	0·70
Fir	0·775
Pitch pine	1·14
Red pine	0·94
Beech	1·09

Example. — To find the proper breadth for a beam of ash, 20 feet long and 12 inches deep, to bear a load of 5000 lbs. on the middle—

---

* This Table is computed from a Table of multiplicands given by Mr. Barlow, to be used with his rules for finding the ultimate transverse strength of different woods. The column of multiplicands in the above Table expresses the *proportionate transverse strengths* of different woods, English oak being taken as unity.

$$\frac{5000 \times 20}{40 \times 1\cdot42 \times 12^2} = 12\cdot2 \text{ inches for the breadth of the beam ; or,}$$

if the breadth were known, 12·2 inches, and the depth sought,
Then,

$$\text{Depth} = \sqrt{\frac{5000 \times 20}{40 \times 1\cdot42 \times 12\cdot2}} = \sqrt{144} = 12 \text{ inches.}$$

230. Mr. Farey's rule is adapted to give results
corresponding to the practice in making the wooden
beams of steam engines, *viz.* to load oak beams at
the rate of 40 lbs. on a bar 1 foot long and 1 inch
square.

Now Mr. Smeaton's experiments on bars of
oak give for the weight that will break a bar
of good English oak 1 inch square and 2 feet
long between the supports, when it is loaded
in the middle - - - - - 280 lbs.
And Mr. Barlow states, that a bar of oak 2
inches square and 7 feet long between the
supports, broke when loaded in the middle
with - - - - - - 637 lbs.

The mean of these two results gives 558½ *lbs. for
the ultimate transverse strength of a bar of oak* 1
foot long between the supports, and 1 *inch square;*
hence Mr. Farey's rule is adapted to make oak beams
of a strength to bear very nearly 14 *times the load
that is to be laid on them* before they would break.
This is a superabundance of strength for beams at
rest, and in using the rule for computing the proper
dimensions for the cross joists of the platform of a suspension bridge, it must be taken with modification :
viz. If they are made to bear three or four times as
much as they will be loaded with, instead of fourteen
times, they will be abundantly strong, and the rule

may therefore be used with 140 for a divisor instead of 40 ; Or by using a divisor = any proportion of $558\frac{1}{2}$ (or say of 560 as a more conveniently divisible number), a corresponding proportion of strength will be computed for the wood.

231. The increase of strength given to wood bearers by trussing them is very great. Sir H. Douglas says, that a trussed beam will bear 10 *times the weight* that it will bear alone ; and that a bridge on oak trusses of 40 feet span, in which none of the timbers exceed 6 inches square, may be made to support artillery. (*See* Sir H. Douglas on Military Bridges.)

And in an experiment made with a portion of the platform of Menai Bridge, it was found that *when trussed* it bore *five times* the weight that bent it when not trussed.*

Strength of Ropes.

232. In the erection of the Menai Bridge some trials were made by Mr. Rhodes of the strength of the ropes used for the hoisting tackle to get up the main chains.

They were as follows : —

                                                     Tons per
                                                     sq. inch.

No. I.   A piece of $5\frac{3}{4}$ inches circumference broke with $6\frac{3}{4}$ tons =   2·56

No. II.   A piece of $4\frac{1}{2}$ inches circumference common laid, broke with $4\frac{1}{10}$th tons =   2·54

No. III. A piece $4\frac{1}{2}$ inches circumference of fine yarn, slack laid, broke with 6 tons =   3·73

* See Provis's Account of Menai Bridge

The mean of Nos. I. and II. may be taken as a standard : *viz.*

That *good rope will break with a strain of* 2·55 *tons per square inch of section.* But it ought not to be strained permanently with more than ⅓d of that, say about ¾ths of a ton. For temporary service, such as a military bridge erected merely for the passage of troops, rope may, perhaps, be loaded with ½ of its breaking strain, or say 1¼ ton per square inch.

Dr. Gregory * gives the following rules for the strength of *hemp rope.*

(The square of the girth, or circumference × 200) = the strain in pounds that it will bear safely.†

Example : — For a rope 4½ inches girth, we have 4½ squared = 20·25 × 200 = 4050 lbs. = 1·12 ton per square inch, or ₁⁵₁ths of the breaking strain in No. II. of Mr. Rhodes's experiments.

For finding the *breaking strain* of ropes, Dr. Gregory gives the following rule : —

$$\frac{\text{The girth square}}{5} = \text{the load } in\ tons \text{ that will break the rope.}‡$$

Example : — A rope 4½ inches girth squared = $\frac{20\cdot25}{5}$ = 4·05 tons. The real breaking strain of a rope 4½ inches girth is by No. II. experiment 4₁⁰ tons.

Another example : — 5¾ girth squared = $\frac{33}{5}$ = 6·6 tons. The real breaking strain in No. III. of Mr. Rhodes's experiments was 6¾ tons. So that Dr. Gregory's rules appear to accord very well with experiment.

* Dr. Gregory's Mathematics for Practical Men, p. 391.

† For *cables*, Dr. Gregory says 120 must be used for a multiplier instead of 200.

‡ Dr. Gregory says this standard is too high for tarred cordage, which is weaker than white cordage.

CONCLUSION.

233. The application of suspension bridges has, within a few years, increased so rapidly, and is still so much on the increase, that it may not be out of place to bestow a few lines on the consideration of when they are, and when they are not, expedient.

234. The prominent quality of a suspension bridge is its independence of the bed of the river that it crosses. Hence it can be thrown across an opening, where it is impracticable, either from rapid current, or from the altitude of the banks, to erect centering for a stone bridge.

Its next most valuable qualities are the facility and expedition with which it can be built, the small amount of materials required, and the consequent economy.

These advantages, added to the elegant lightness of suspension bridges, have combined to throw a degree of charm about them, which is perhaps becoming exaggerated, and may lead to their adoption in unfitting situations.

235. It should be remembered, that while suspension bridges are built on the proportions hitherto adopted, even in the strongest, they are incomparably slighter than stone or cast-iron arch bridges. There is no suspension bridge in existence that would be fit to bear permanently the load that is daily and hourly crowded on London Bridge.

A bridge destined to be a great and perpetual thoroughfare, exposed not only to be frequently quite filled with people, and to the passage of troops, but also to the rapid motion of great numbers of heavy vehicles ; in fine, a bridge in a busy part of a great city ought not to be on the suspension principle.

238. For if it were made no stronger than our strongest suspension bridges, it would not possess sufficient stability. If, on the other hand, the strength were increased to a sufficient extent to enable it to bear safely its constant work, the weight, the difficulty of getting up the chains, and the increase in the masonry part, would so raise the expense, that it is doubtful how far it could be brought under that of a stone or cast iron bridge.

Add to which, a suspension bridge would never equal in stability a common arch bridge, because it is subject to vibrations, the law of which is not sufficiently known to calculate their precise results in practice, but which certainly are more dangerous in a heavy bridge than in a light one. The object, therefore, in building a suspension bridge is, either to make it so light that its own vibration shall not hurt it; or if, as in nine cases out of ten, that cannot be done, then to make it so heavy and stiff, in proportion to the load it will have to carry, that the load shall not cause it to vibrate much. This, for a bridge liable to be constantly loaded with as much as it could contain, would be impracticable.

239. For large openings, where it is of importance to have a permanent passage, and yet where the number of passengers is seldom great at a time, suspension bridges are admirably fitted, because they can be carried to almost any span, and any height, for a comparatively moderate expense.

There are also multitudes of situations, where it has been usual to build arch bridges of stone at great

P

expense, and where the traffic is not at all beyond the measure of strength that may judiciously be given to a suspension bridge.

240. For military bridges they are well fitted; the chains or cables, the platform, and even timbers ready prepared to frame suspension piers. — An entire suspension bridge, in fact, — might be carried more conveniently than a pontoon bridge, and could be rigged up for use in very little time.

They would be also peculiarly well adapted for crossing chasms in mountainous countries. On the Simplon and St. Gothard roads, for instance, the celebrated passes from Switzerland to Italy, the chasms that have to be crossed by bridges are frequently many hundred feet in depth, although not broad, and the faces of the rock so perpendicular, or overhanging, as to give hardly any means of erecting centering for an arch bridge. The expense, consequently, of making them, must have been very great. A suspension bridge, moreover, on a great military pass, would give the inhabitants greater command over it; for by knocking out a few connecting bolts, a whole bridge might be dismantled very rapidly, without being destroyed, to check or retard the enemy's passage; whereas, to cut off the passage of a stone bridge, it must be blown up, and cannot be renewed but with great expense and loss of time.

241. For piers, or jetties on the sea coast, they appear to be peculiarly adapted, from the openness of their construction. If the suspension towers are founded on piles, and themselves made of strong but open framework, and if the chains and platform are properly combined to get as much stiffness with as little weight as possible, so that they may resist vi-

bration, without being so heavy as to be endangered by the vibration they cannot resist, a suspension pier may be buried in the waves without being hurt.

242. As to the durability of suspension bridges, nothing but time can determine it. The chains are tried with nine tons per square inch, and do not stretch with that strain, according to which they are proportioned. But it does not follow, that the iron may not receive injury from a permanent load of nine tons per square inch, although it receives no injury from it during the short time it is under proof. It would be a useful experiment to strain bars of iron with different loads from seven tons per square inch upwards, and to leave them under their loads for several years exposed to the air and damp; in fact, under the same circumstances as in a bridge. Again, as to preserving the chains from rust by varnishing and painting, they are certainly protected by it in some degree; nevertheless, destruction does go on, as is evident by the necessity of scraping and repainting every few years; and to what extent that silent corrosion may eat into the fibre, and injure the tenacity of the iron without such injury being discernible to the eye, is not determined.

It might, perhaps, be well, while these points remain unsettled by experimental knowledge, to give to suspension bridges more strength than is usual, or else to take out a bar or a bolt every few years, and prove it over again to ascertain when the bridge ought to be repaired. For it does not at all follow, that a chain bridge that will bear 1000 tons now, will bear 1000 tons a century hence.

THE END.

London :
Printed by A. & R. Spottiswoode,
New-Street-Square.

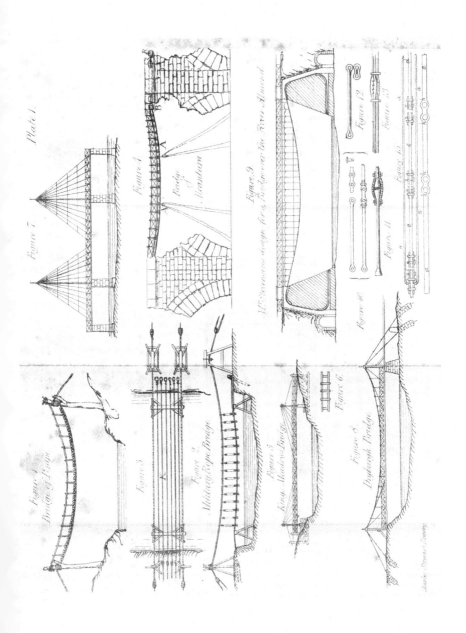

The material originally positioned here is too large for reproduction in this reissue. A PDF can be downloaded from the web address given on page iv of this book, by clicking on 'Resources Available'.

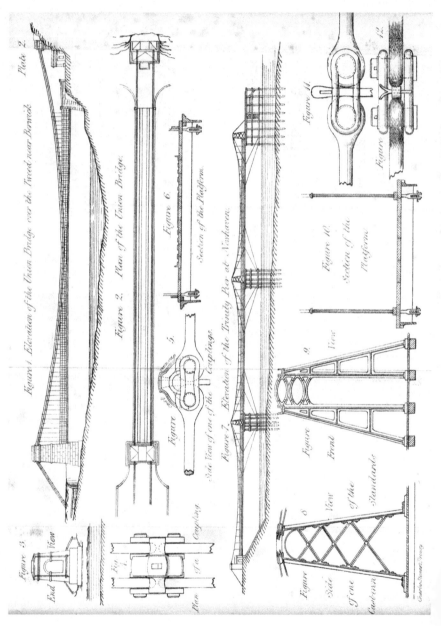

Plate 2.

Figure 1 Elevation of the Union Bridge over the Tweed near Berwick.

Figure 2. Plan of the Union Bridge.

Figure 3. End View

Figure 4.

Figure 5.

Figure 6. Section of the Platform.

Side View of one of the Couplings.

Figure. Plan of the Coupling

Figure 7. Elevation of the Trinity Pier at Newhaven.

Figure 10. Section of the Platform

Figure 9. Front View

Figure 8. Side View of the Standards.

The material originally positioned here is too large for reproduction in this reissue. A PDF can be downloaded from the web address given on page iv of this book, by clicking on 'Resources Available'.

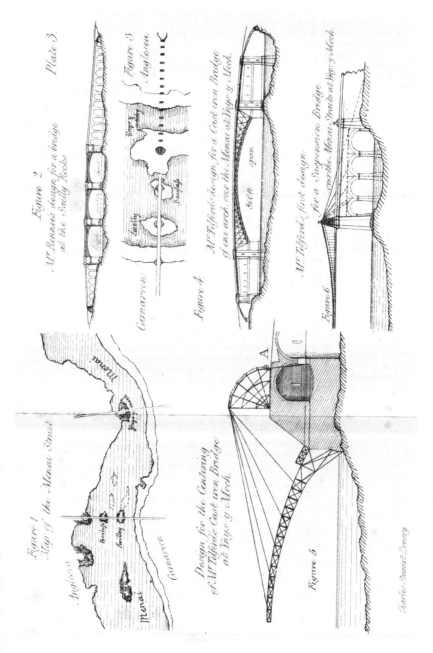

The material originally positioned here is too large for reproduction in this reissue. A PDF can be downloaded from the web address given on page iv of this book, by clicking on 'Resources Available'.

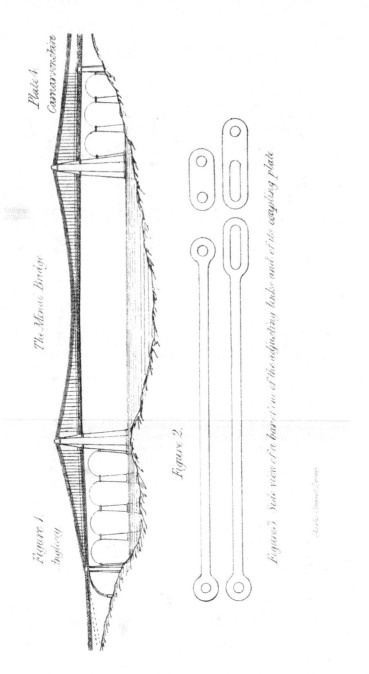

Plate 4.
Carnarvonshire

The Menai Bridge

Figure 1.
Anglesey

Figure 2.

Figure 3. Side view of a bar view of the adjusting links and of its coupling plate

The material originally positioned here is too large for reproduction in this reissue. A PDF can be downloaded from the web address given on page iv of this book, by clicking on 'Resources Available'.

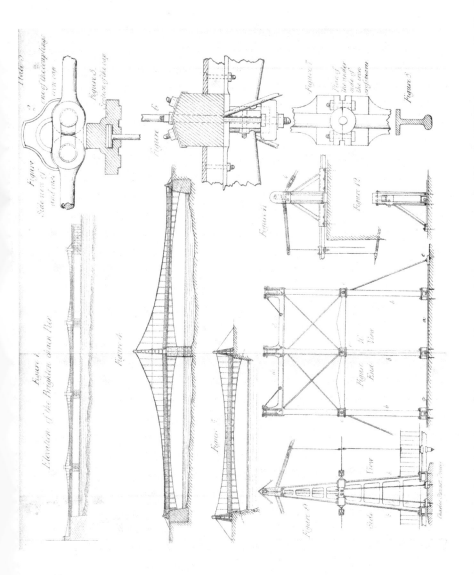

The material originally positioned here is too large for reproduction in this reissue. A PDF can be downloaded from the web address given on page iv of this book, by clicking on 'Resources Available'.

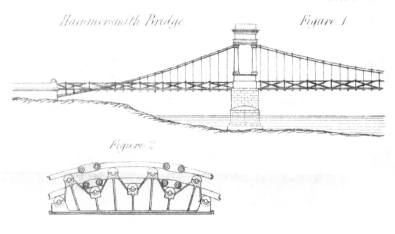

Plate 6.

Hammersmith Bridge.

Figure 1

Figure 2

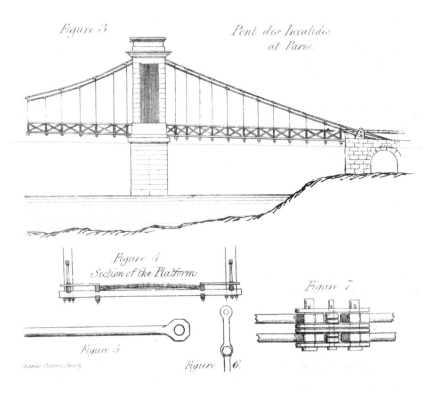

Figure 3

Pont des Invalides at Paris.

Figure 4
Section of the Platform

Figure 7.

Figure 5

Figure 6.

The material originally positioned here is too large for reproduction in this reissue. A PDF can be downloaded from the web address given on page iv of this book, by clicking on 'Resources Available'.

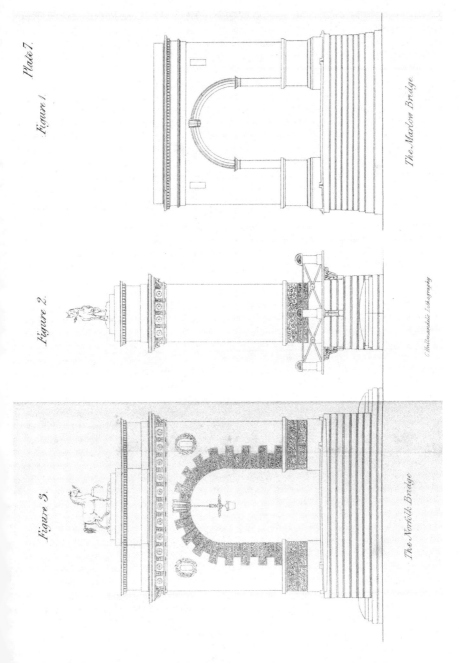

The material originally positioned here is too large for reproduction in this reissue. A PDF can be downloaded from the web address given on page iv of this book, by clicking on 'Resources Available'.

Printed in the United States
By Bookmasters